INTERNATIONAL CENTRE FOR MECHANICAL SCIENCES

EBERHARD BROMMUNDT
TECHNICAL UNIVERSITY OF DARMSTADT

VIBRATIONS OF CONTINUOUS SYSTEMS

THEORY AND APPLICATIONS

COURSE HELD AT THE DEPARTMENT
FOR MECHANICS OF DEFORMABLE BODIES
SEPTEMBER - OCTOBER 1969

UDINE 1969

COURSES AND LECTURES - N. 1

ISBN 978-3-211-81305-8 ISBN 978-3-7091-2918-0 (eBook)
DOI 10.1007/978-3-7091-2918-0

Originally published by Springer Vienna in 1970

First Reprint.

P R E F A C E

This booklet contains the notes of my lectures on vibrations of (solid) continuous systems delivered at CISM in Fall of 1969. The lectures were presented to an auditory of engineers and physicists interested in various branches of mechanics.

Starting from vibrations of conservative, linear systems I tried to give an introduction to some problems, methods of solution, and phenomena of non-conservative and nonlinear systems. The examples chosen to demonstrate the different notions and procedures are very simple to avoid lengthy calculations which might hide the basic ideas.

I would like to express my sincere thanks to the authorities of CISM, in particular to professors W.Olszak and L.Sobrero, for their kind invitation and continued interest.

E.Brommundt

Udine, October 1969.

0. Introduction

Objectives of vibrational investigations are technical (physical, chemical), biological, economic, etc. systems.

Purposes of such investigations :

to "comprehend" phenomena observed (experimentally) in actual systems;

to "predict" the behavior, qualitatively as well as quantitatively, of systems not yet (experimentally) tested,and of systems which are only projected as in engineering design.

The procedures of these investigations are always similar, see Fig. O. 1.

There is no way to compare mathematically the results obtained for the model with the behavior of the real system.

In these lectures we shall restrict ourselves to

(ct. p. 8)

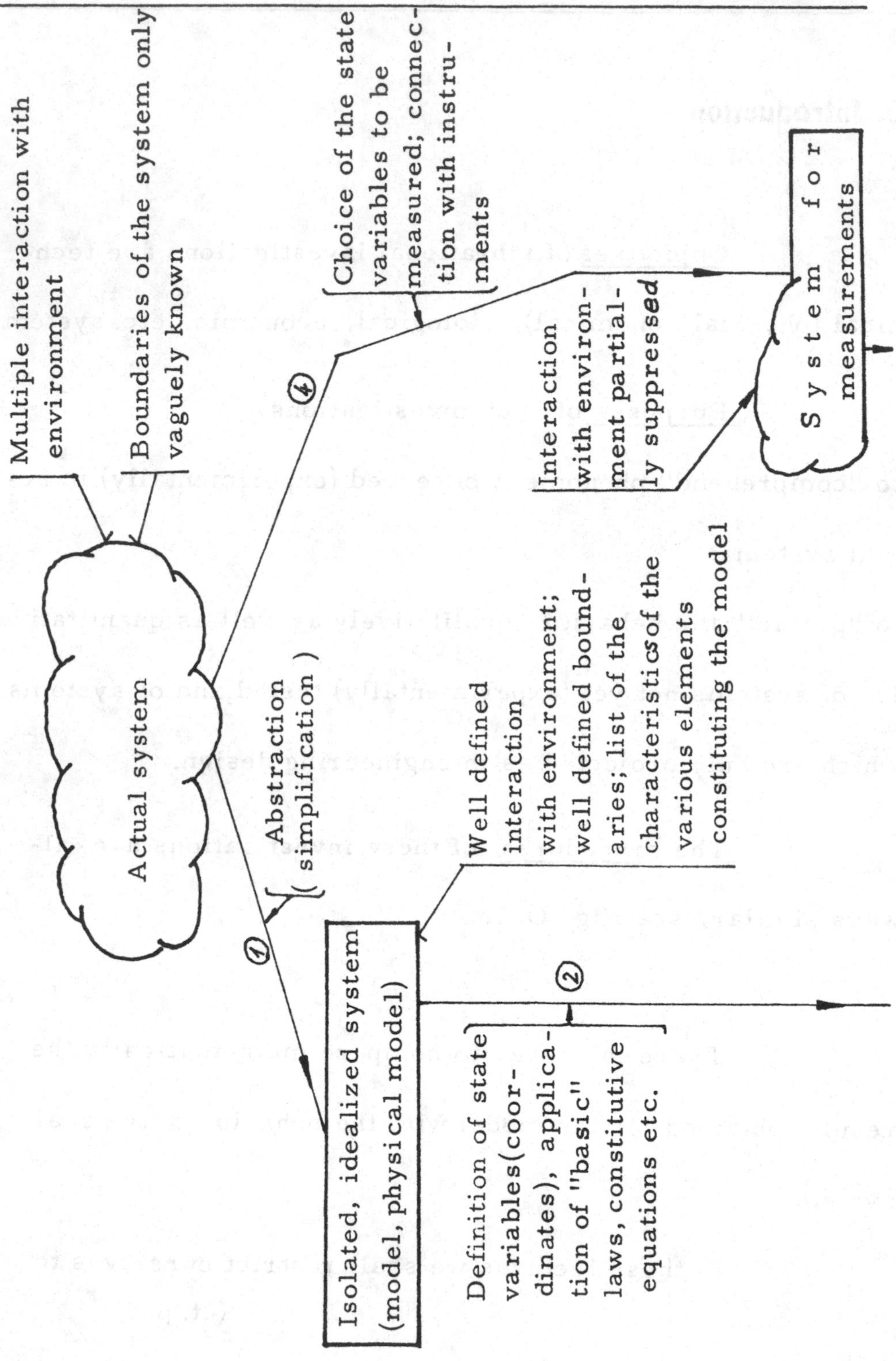
Multiple interaction with environment
Boundaries of the system only vaguely known
Actual system
Abstraction (simplification)
1
Isolated, idealized system (model, physical model)
Well defined interaction with environment; well defined boundaries; list of the characteristics of the various elements constituting the model
Definition of state variables (coordinates), application of "basic" laws, constitutive equations etc.
2
4
Choice of the state variables to be measured; connection with instruments
Interaction with environment partially suppressed
System for measurements

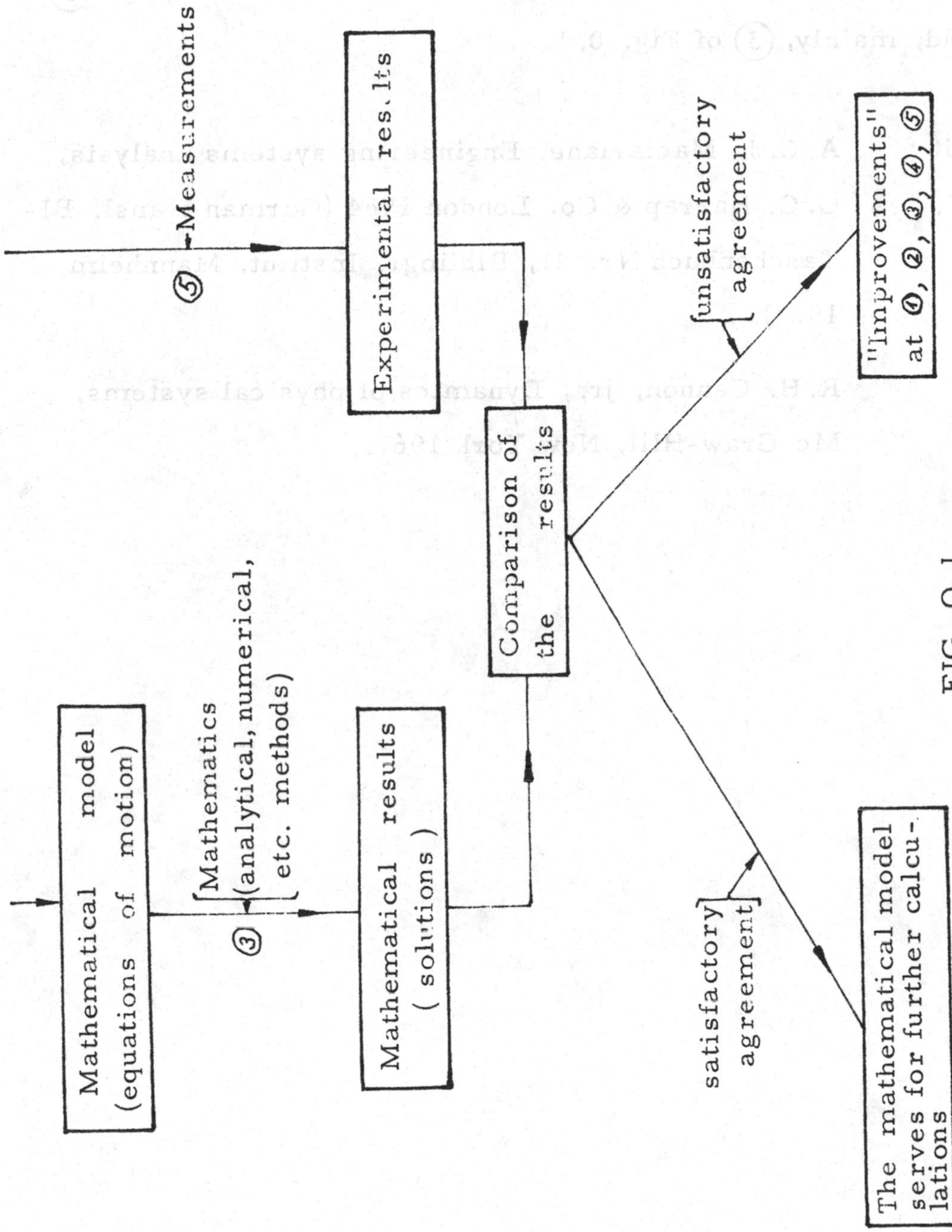
Mathematical model (equations of motion)
③ Mathematics (analytical, numerical, etc. methods)
Mathematical results (solutions)
⑤ Measurements
Experimental results
Comparison of the results
unsatisfactory agreement
"Improvements" at ①, ②, ③, ④, ⑤
satisfactory agreement
The mathematical model serves for further calculations

FIG. O.1

mechanical system. We shall be concerned with the steps (2) and, mainly, (3) of Fig. 0.1.

Lit : A. G. J. Macfarlane, Engineering systems analysis, G. G. Harrap & Co, London 1964 (German transl. BI-Taschenbuch Nr. 81, Bibliogr. Institut, Mannheim 1967).

R. H. Cannon, jr., Dynamics of physical systems, Mc Graw-Hill, New York 1967.

1. Continuous systems

1. 1 Coordinates

1. 11 Reference coordinates

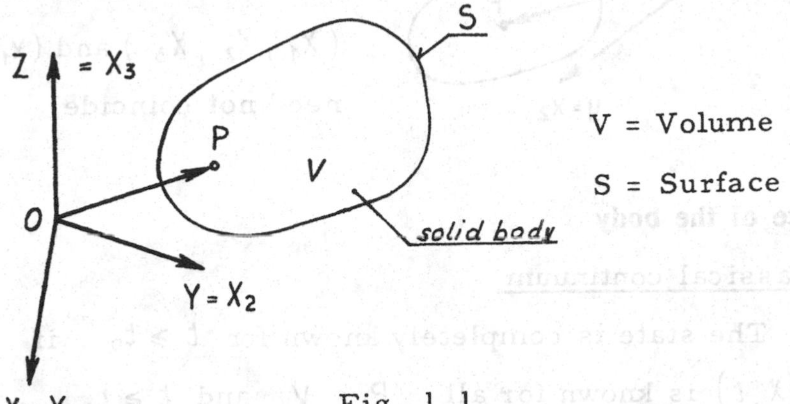

Fig. 1. 1

The particle P is denoted by the Lagrangian coordinates

$$(X, Y, Z) = (X_1, X_2, X_3) = \underset{\sim}{X}$$

which may be interpreted as its (curvilinear) coordinates in a certain reference configuration of the body. (Unique notation, continuously differentiable ⟶ no cracks, etc.)

1. 12 Position (spatial, Eulerian) coordinates

The position of the particle P at the time t is given by

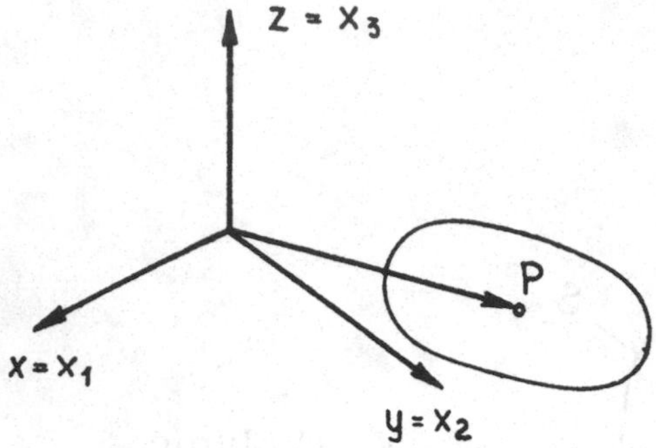

$$(x(\underset{\sim}{X}.t), y(\underset{\sim}{X}.t), z(\underset{\sim}{X}.t)) = (x_1(\underset{\sim}{X}.t), x_2(\underset{\sim}{X}.t), x_3(\underset{\sim}{X}.t)) = \underset{\sim}{x}(\underset{\sim}{X}.t)$$

The systems of coordinates (X_1, X_2, X_3) and (x_1, x_2, x_3) need not coincide.

1. 2 State of the body

1. 21 Classical continuum

The state is completely known for $t \geq t_0$ if $\underset{\sim}{x} = \underset{\sim}{x}(\underset{\sim}{X}, t)$ is known for all $P \in V$ and $t \geq t_0$.

1. 22 Modern extensions

1. 221 Thermodynamical state variables are taken into account.

1. 222 Microstructure of the material (grains, crystals, complex molecules) are taken into account by associating "directions" etc. with the particles ⟶ multipolar media.

From a formal point of view both extensions mean that the number of the state variables $x_i(\underset{\sim}{X}.t)$ is increased,

$$i = 1, \ldots, N; \quad N > 3; \quad (x_1(\underset{\sim}{X}.t), \ldots, x_N(\underset{\sim}{X}.t)) = \underset{\sim}{x}(\underset{\sim}{X}.t)$$

Lit. : C. Truesdell, The elements of continuum mechanics, Springer, Berlin 1966

A. C. Eringen, Mechanics of continua, J. Wiley, New York 1967

W. Jaunzemis, Continuum mechanics, Macmillan, New York 1967

1. 3 Problem

Find $\underset{\sim}{x} = \underset{\sim}{x}(\underset{\sim}{X}, t)$ and, maybe, some quantities derived from $\underset{\sim}{x}$; e. g., stresses, strains etc.

2. Some classical, conservative, linear systems

2. 1 Longitudinal vibrations of a rod

2. 11 The mechanical model

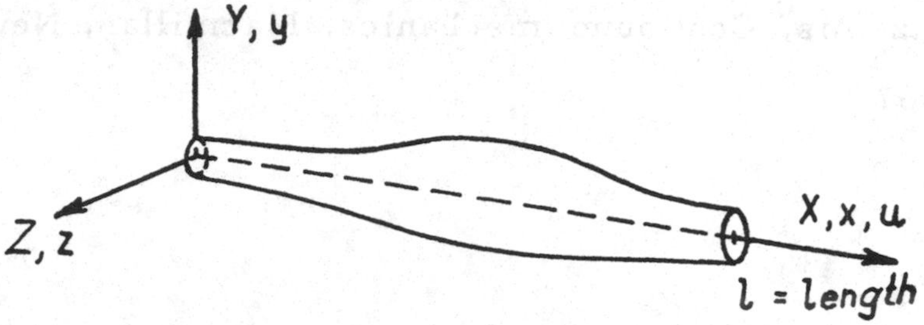

Fig. 2. 1

Rod originally straight, reference configuration as shown in Fig. 2. 1

Problem : Investigate the longitudinal vibrations.

Simplifying assumption : All originally plane cross-sectional areas X = const. remain plane and perpendicular to the X- axis.

Displacement : $u(\underset{\sim}{X}, t) = u(X, t) = x(X, t) - X.$

2.12 Equation of motion

2.121 Deformation

Strain : $\varepsilon = \dfrac{\partial u}{\partial X}$

2.122 Constitutive equation

Hooke's law $\sigma = E\varepsilon$

σ - stress, $E = E(X)$ - modulus of elasticity

2.123 Force

$$F = A \cdot \sigma$$

$A(X)$ cross-sectional area

2.124 Equilibrium

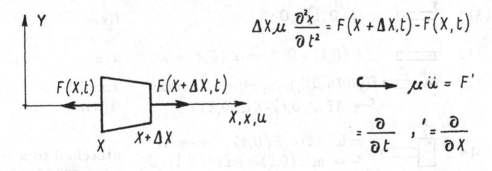

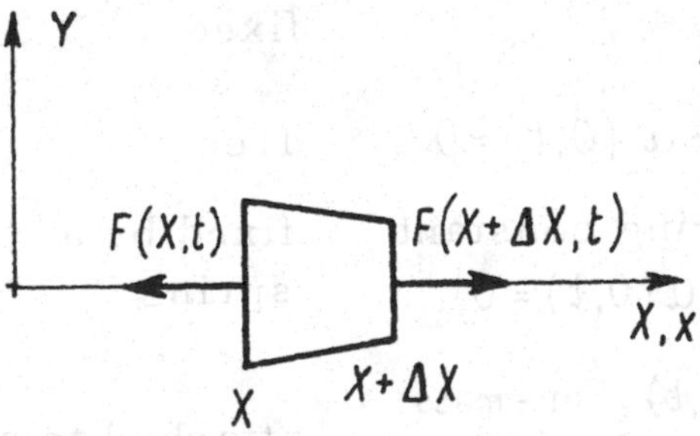

$$\Delta X \mu \frac{\partial^2 x}{\partial t^2} = F(X+\Delta X, t) - F(X,t)$$

$$\longrightarrow \mu \ddot{u} = F'$$

$$\dot{} = \frac{\partial}{\partial t}, \quad ' = \frac{\partial}{\partial X}$$

$\mu = \rho A$ - mass density per unit length

ρ - mass density

2.125 Equation of motion

Elimination of F yields

$$\mu \ddot{u} - \left[A E u' \right]' = 0$$

$\mu \, , \, A \, , \, E > 0$, sufficiently smooth

2.13 Boundary conditions

2.131 Homogeneous boundary conditions

Boundary conditions for $X = 0$: End:

2.1311 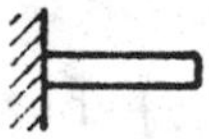 $u(0,t) = 0$ fixed

2.1312  $F(0,t) = 0 \hookrightarrow u'(0,t) = 0$ free

2.1313 

$F(0,t) = K u(0,t)$, k-spring constant fixed by a spring

$\hookrightarrow AEu'(0,t) - Ku(0,t) = 0$

2.1314 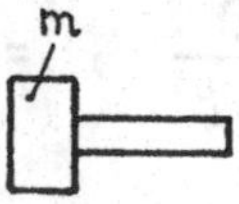

$m \ddot{u}(0,t) = F(0,t)$, m-mass attached to a rigid mass

$\hookrightarrow m\ddot{u}(0,t) - AEu'(0,t) = 0$

Similar conditions hold for $X = l$.

2.132 Nonhomogeneous boundary condition

2.1321 $F(l,t) = P(t)$, $P(t)$ given force

$\hookrightarrow AE\,u'(l,t) = P(t)$

2.1322 $u(l,t) = f(t)$, $f(t)$ given displacement

2.2 Further conservative, linear systems

2.21 String

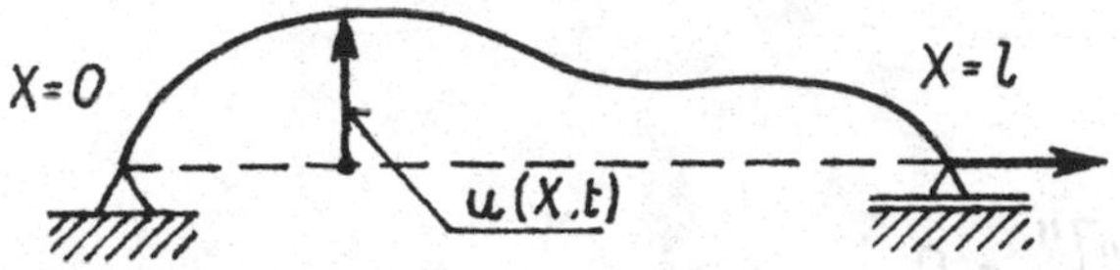

P - constant force

$\mu(X)$ - mass per unit length

2.211 Equation of motion

$$\mu \ddot{u} - P u'' = 0$$

2.212 Boundary conditions (homogeneous)

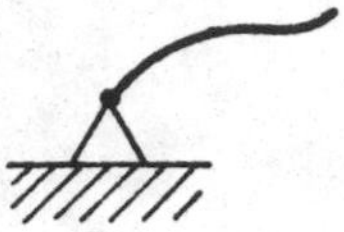

$u(0,t) = 0$ fixed

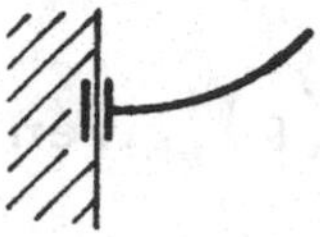

$u'(0,t) = 0$ sliding

etc.

2.22 Euler-Bernoulli beam

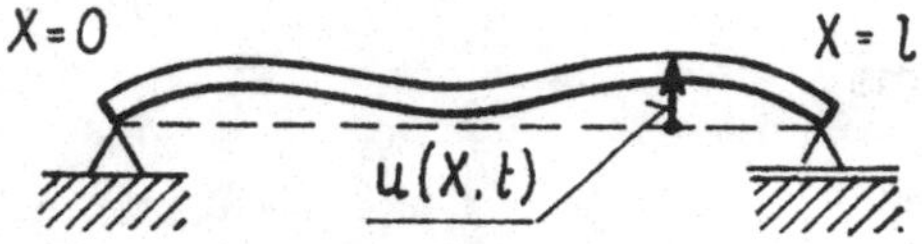

EI - bending stiffness

μ - mass per unit length (may depend on X)

2.221 Equation of motion

$$\mu \ddot{u} + \left[EI\, u'' \right]'' = 0$$

2.222 Boundary conditions (homogeneous)

2.2221 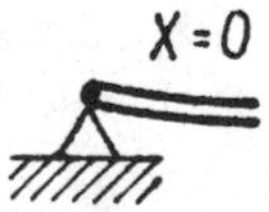

$u(0,t) = 0, \; u''(0,t) = 0$ supported

2.2222 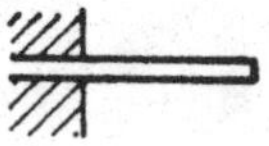

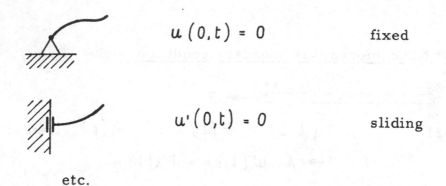

$u(0,t) = 0, \; u'(0,t) = 0$ clamped

2.2223  $u''(0,t) = 0\,,\quad \left[EI\,u''\right]'\Big|_{X=0} = 0$ free

2.2224 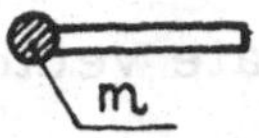

$\begin{cases} u''(0,t) = 0\,, \\ m\ddot{u}(0,t) + \left[EI\,u''\right]'\Big|_{X=0} = 0 \end{cases}$ point mass at $X = 0$

2.23 Timoshenko beam

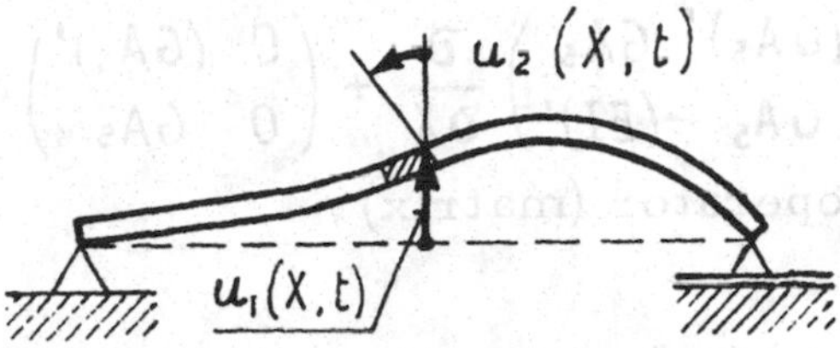

u_1 – lateral displacement
u_2 – angular displacement
EI – bending diffness
GA_s – shear stiffness
μ_1 – mass per unit length
μ_2 – rotatory inertia per unit length

} may depend on X

2.231 Equation of motion

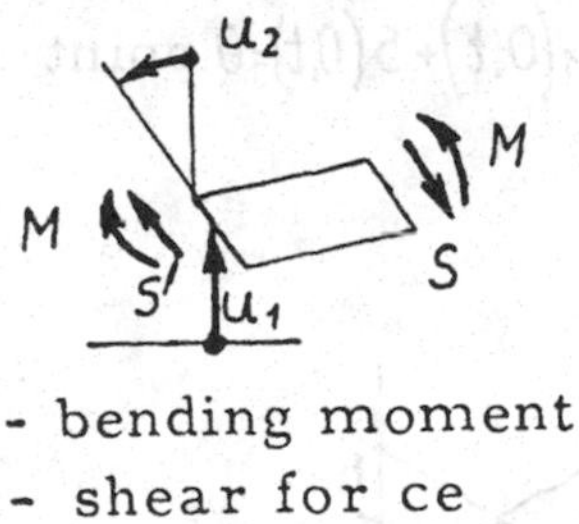

M - bending moment
S - shear for ce

$$\mu_1 \ddot{u}_1 = -S'$$

$$\mu_2 \ddot{u}_2 = M' - S$$

$$EI\,u_2' = M$$

$$GA_s(u_2 - u_1') = S$$

$$\hookrightarrow \left|\begin{array}{l} \mu_1 \ddot{u}_1 + \left[GA_s(u_2 - u_1')\right]' = 0 \\ \mu_2 \ddot{u}_2 + GA_s(u_2 - u_1') - (EI\,u_2')' = 0 \end{array}\right.$$

Matrix notation

$$\underset{\sim}{\mu}\,\ddot{\underset{\sim}{u}} + \underset{\sim}{L}\,\underset{\sim}{u} = 0$$

where $\underset{\sim}{u} = \begin{pmatrix} u_1 \\ u_2 \end{pmatrix}$ - state vector

$\underset{\sim}{\mu} = \begin{pmatrix} \mu_1 & 0 \\ 0 & \mu_2 \end{pmatrix}$ - inertia matrix

$$\underset{\sim}{L} = \begin{pmatrix} -GA_s & 0 \\ 0 & -EI \end{pmatrix} \frac{\partial^2}{\partial X^2} + \begin{pmatrix} -(GA_s)' & GA_s \\ -GA_s & -(EI)' \end{pmatrix} \frac{\partial}{\partial X} + \begin{pmatrix} 0 & (GA_s)' \\ 0 & GA_s \end{pmatrix}$$

- linear differential operator (matrix)

2.232 Boundary conditions

2.2321 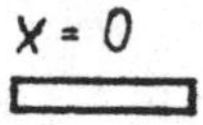

$M(0,t) = 0$, $S(0,t) = 0$ free

2.2322 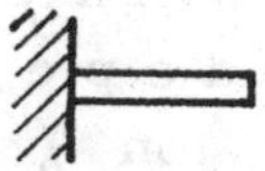 $u_1(0,t) = 0$, $u_2(0,t) = 0$ fixed

2.2323 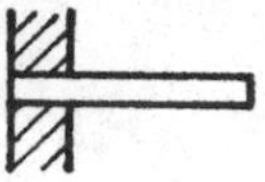 $u_1(0,t) = 0$, $u_1'(0,t) = 0$ guided

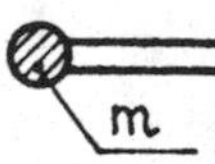

$M(0,t) = 0$, $m\ddot{u}_1(0,t) + S(0,t) = 0$ point mass

2.24 Plate, transverse vibrations

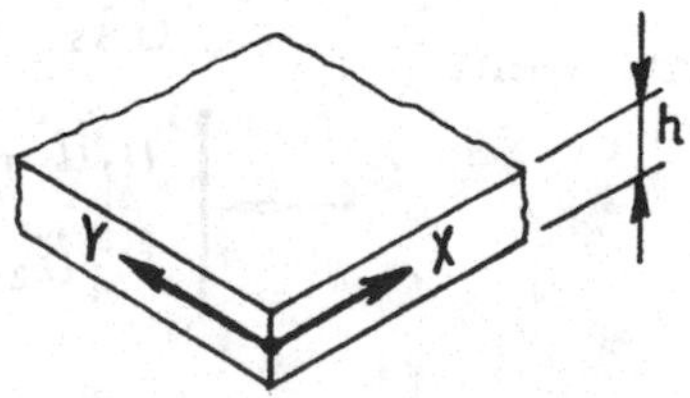

h - thickness (uniform)

u - deflection

μ - mass per unit area, $\mu = \varrho h$

$K = Eh^3/12(1-\nu^2)$ bending stiffness

ν - Poisson's ratio

2.241 Equation of motion

$$\mu \ddot{u} + K \nabla^2 \nabla^2 u = 0$$

$$\nabla^2 = \text{div grad} = \frac{\partial^2}{\partial X^2} + \frac{\partial^2}{\partial Y^2} \quad - \quad \text{Laplacian operator}$$

2.242 Boundary conditions

For $X = 0$; Y, t arbitrary

2.2421 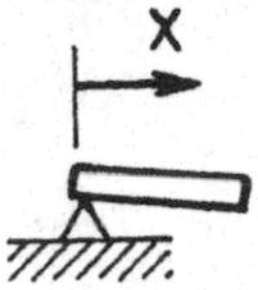

$$u = 0, \quad \frac{\partial^2 u}{\partial X^2} + \nu \frac{\partial^2 u}{\partial Y^2} = 0$$

supported

2.2422 

$$u = 0, \quad \frac{\partial u}{\partial X} = 0$$

clamped

2.2423 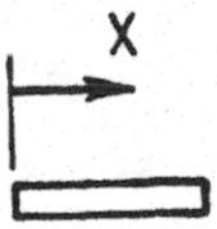

$$\frac{\partial^2 u}{\partial X^2} + \nu \frac{\partial^2 u}{\partial Y^2} = 0,$$

free

$$\frac{\partial}{\partial X}\left[\frac{\partial^2 u}{\partial X^2} + (2-\nu)\frac{\partial^2 u}{\partial Y^2}\right] = 0$$

Lit. : W. Flügge, Handbook of engineering mechanics, McGraw-Hill, New York 1962, Sect. 61

2.25 Isotropic continuous body (three-dimensional)

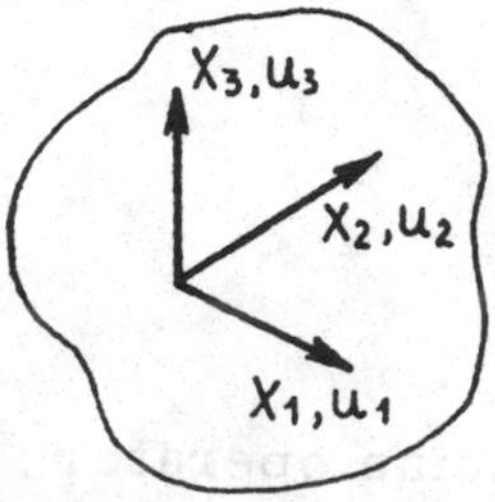

$\underset{\sim}{u} = \{u_1, u_2, u_3\}$ displacement vector
ϱ - density
ν - Poisson's ratio
G - shear modulus

2.251 Equation of motion

$$\frac{G}{1-2\nu} \, grad\, div\, \underset{\sim}{u} + G \nabla^2 \underset{\sim}{u} - \varrho \ddot{\underset{\sim}{u}} = 0$$

$$div\, \underset{\sim}{u} = \frac{\partial u_1}{\partial X_1} + \frac{\partial u_2}{\partial X_2} + \frac{\partial u_3}{\partial X_3} = e \text{ - } \quad \text{dilatation}$$

$$grad\, div\, \underset{\sim}{u} = \nabla^2 \underset{\sim}{u} + curl\, curl\, \underset{\sim}{u}.$$

Special motions

$curl\, \underset{\sim}{u} = 0$: $\quad c_{dil}^2 \nabla^2 \underset{\sim}{u} - \ddot{\underset{\sim}{u}} = 0$ only dilatation (no distortion)

$$c_{dil}^2 = 2\,(1-\nu)\, G/(1-2\nu)\, \varrho\, .$$

$\operatorname{div} \underset{\sim}{u} = 0$: $\quad c_{dist}^2 \nabla^2 \underset{\sim}{u} - \ddot{\underset{\sim}{u}} = 0$ only distortion (no dilatation)

$$c_{dist}^2 = G/\varrho$$

2.252 Boundary conditions

Zero stresses, zero displacements etc., cf. Flügge, Handbook (cited above).

3. Wave solutions

3. I Longitudinal waves in an uniform rod (cf. 2. 1)

μ , A , E - constant

Equation of motion (cf. 2. 125)

$$c^2 u'' - \ddot{u} = 0 \quad , \quad c^2 = AE/\mu \qquad (*)$$

3. 11 Travelling waves in an infinite rod

General solution of (*) :

$$u = f(x - ct) + g(x + c,t); \; f,g \text{ arbitrary functions sufficiently smooth}$$

f and g represent travelling waves

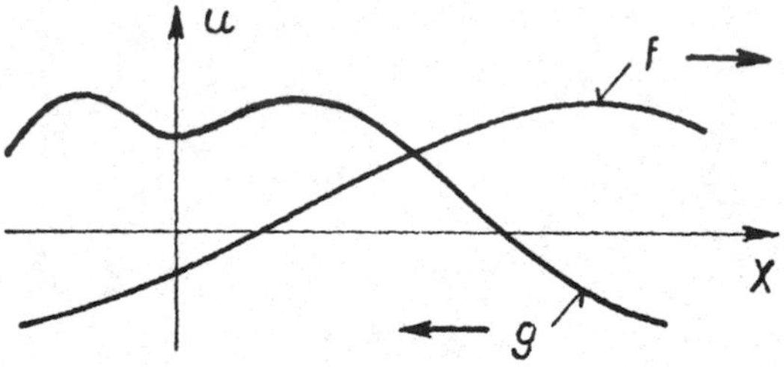

Fig. 3. 1

The arrows indicate the direction of the propagation,

c - wave velocity.

No distortion (dispersion) of the waves.

3.111 Initial value problem

Given : $u(X,0) = \varphi(X)\ ,\ u(X,0) = \psi(X)$;

find f and g

D'Alembert's solution :

$$\left.\begin{aligned} f(\xi) + g(\xi) &= \varphi(\xi) \\ -cf'(\xi) + cg'(\xi) &= \psi(\xi) \end{aligned}\right\} \rightarrow \left\{\begin{aligned} f(\xi) &= \frac{1}{2}\varphi(\xi) - \frac{1}{2c}\int^{\xi} \psi(\zeta)\, d\zeta \\ g(\xi) &= \frac{1}{2}\varphi(\xi) + \frac{1}{2c}\int^{\xi} \psi(\zeta)\, d\zeta \end{aligned}\right.$$

$$u(X,t) = \frac{1}{2}\left[\varphi(X-ct) + \varphi(X+ct)\right] + \frac{1}{2c}\int_{X-ct}^{X+ct} \psi(\zeta)\, d\zeta$$

3.12 Waves in a finite rod

Equation of motion :

$$c^2 u'' - \ddot{u} = 0 \qquad (*)$$

Chosen boundary conditions

$$\left.\begin{aligned} u(0,t) &= 0 \qquad \text{fixed} \\ u'(l,t) &= 0 \qquad \text{free} \end{aligned}\right\} \qquad (**)$$

Initial conditions

$$\left.\begin{aligned} u(X,0) &= \varphi(X) \\ \dot{u}(X,0) &= \psi(X) \end{aligned}\right\} \text{ for } 0 \leq X \leq l \qquad (***)$$

(***) satisfy (**).

General solution of (*), cf. 3.11 :

$$u = f(X - ct) + g(X + ct).$$

From (**)

$$\left.\begin{aligned} f(-\xi) + g(\xi) &= 0 \\ -f'(l-\xi) + g'(l+\xi) &= 0 \end{aligned}\right\} \text{(A)}$$

' = derivative with resp. to the argument

From (***)

$$\left.\begin{aligned} f(\xi) + g(\xi) &= \varphi(\xi) \\ -cf'(\xi) + cg'(\xi) &= \psi(\xi) \end{aligned}\right\} \text{(B)}$$

From (B) we obtain

$$\left.\begin{aligned} f(\xi) &= \frac{1}{2}\varphi(\xi) - \frac{1}{2c}\int_0^\xi \psi(\zeta)\,d\zeta \\ g(\xi) &= \frac{1}{2}\varphi(\xi) + \frac{1}{2c}\int_0^\xi \psi(\zeta)\,d\zeta \end{aligned}\right\} \text{for } 0 \le \xi \le l$$

Continuation by means of the equations (A) :

$$g(l+\xi) = f(l-\xi) + g(l) - f(l) \quad \text{reflection at } X = l,$$

$$f(-\xi) = -g(\xi) \quad \text{reflection at } X = 0.$$

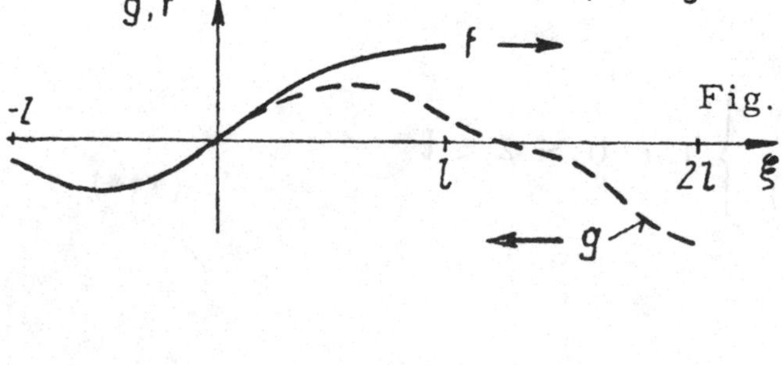

Fig. 3.2

3.2 Waves in an infinite, uniform Timoshenko beam

cf. 2.23; μ_1, μ_2, EI, GA_s - constant.

Equations of motion (matrix notation, cf. sect. 2.231)

$$\underset{\sim}{\mu}\,\ddot{\underset{\sim}{u}} - \begin{pmatrix} GA_s & 0 \\ 0 & EI \end{pmatrix} \underset{\sim}{u}'' + \begin{pmatrix} 0 & GA_s \\ -GA_s & 0 \end{pmatrix} \underset{\sim}{u}' + \begin{pmatrix} 0 & 0 \\ 0 & GA_s \end{pmatrix} \underset{\sim}{u} = 0. \quad (*)$$

No solutions $\underset{\sim}{u} = \underset{\sim}{f}(X \pm ct)$ with arbitrary f

3.21 Special waves

$$\text{Assumption:} \quad \underset{\sim}{u} = \underset{\sim}{a}\, e^{i\varkappa(X-ct)}, \quad i = \sqrt{-1}. \quad (**)$$

(**) in (*):

$$\left\{ -\underset{\sim}{\mu}\,\varkappa^2 c^2 + \begin{pmatrix} GA_s & 0 \\ 0 & EI \end{pmatrix} \varkappa^2 + \begin{pmatrix} 0 & GA_s \\ -GA_s & 0 \end{pmatrix} i\varkappa + \begin{pmatrix} 0 & 0 \\ 0 & GA_s \end{pmatrix} \right\} \underset{\sim}{a} = 0.$$

$\leftrightarrow$ homogeneous system of equations for $\underset{\sim}{a} = \{a_1, a_2\}$.

$$0 = \Delta(\varkappa, c) = \det\{\ldots\} = \begin{vmatrix} -\mu_1 \varkappa^2 c^2 + GA_s \varkappa^2 & GA_s i\varkappa \\ -GA_s i\varkappa & -\mu_2 \varkappa^2 c^2 + EI\varkappa^2 + GA_s \end{vmatrix}$$

$$c^2_{I/II} = \frac{1}{2}\left\{ \frac{GA_s}{\mu_1} + \frac{EI}{\mu_2} + \frac{GA_s}{\mu_2 \varkappa^2} \right\} \pm \sqrt{\frac{1}{4}\left\{\ldots\right\}^2 - \frac{EI}{\mu_2}\frac{GA_s}{\mu_1}}$$

Wave length $\lambda = 2\pi/\varkappa$; two waves,

dispersion: $c_{I/II}$ depend on $\varkappa$.

$$\varkappa \to \infty: \quad c_{I\infty}^2 = \frac{GA_s}{\mu_1} = c_s^2 \quad \text{- shear wave}$$

$$c_{II\infty}^2 = \frac{EI}{\mu_2} = c_B^2 \quad \text{-bending weve}$$

} cf. Flügge below

Euler - Bernoulli beam : $\mu_2 \to 0$, $c_B \to \infty$

($\triangleq$ parabolic diff. equ.)

Lit. : W. Flügge, Die Ausbreitung von Biegungswellen in Stäben, ZAMM 22 (1942) 312-318

3. 3 Waves in a three-dimensional continuum

3. 31 Infinite continuum

Equations of motion cf. 2. 251

$$\text{curl}\ \underset{\sim}{u} = 0: \quad c_{dil}^2 \nabla^2 \underset{\sim}{u} - \ddot{u} = 0; \quad c_{dil}^2 = 2(1-\nu)G/(1-2\nu)\varrho$$

waves of dilatation

plane wave - longitudinal, no dispersion

$$\operatorname{div} \underset{\sim}{u} = 0 : \quad c^2_{dist.} \nabla^2 \underset{\sim}{u} - \ddot{u} = 0 \quad ; \; c^2_{dist.} = G / \rho$$

waves of distortion

plane wave-transversal, no dispersion

3. 32 Bounded continuum

3. 321 Half space

Rayleigh surface waves; no dispersion

cf. A. E. Love, A treatise on the mathematical theory of elasticity, Dover Publicat., New York 1944; Sect 64 (by E. E. Zajac) in Flügge's Handbook.

3. 322 Vibrations of a circular cylinder

Pochchammer, Chree waves;

Torsional, longitudinal and transversal vibrations;

cf. Love and Zajac cited above.

4. Other forms of the equations of motion for linear, finite systems

4. 1 Variational problem

Example : Euler-Bernoulli-beam

Notation see sect. 2. 22

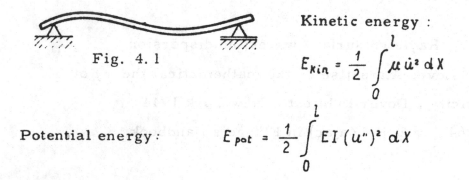

Fig. 4. 1

Kinetic energy :

$$E_{Kin} = \frac{1}{2}\int_0^l \mu \dot{u}^2 \, dX$$

Potential energy:

$$E_{pot} = \frac{1}{2}\int_0^l EI\,(u'')^2 \, dX$$

Hamilton's principle

$$\int_{t_1}^{t_2} (E_{Kin} - E_{pot}) \, dt = \text{extremum} \,, \text{ or}$$

$$\delta \int_{t_1}^{t_2} (E_{Kin} - E_{pot}) \, dt = 0 \qquad (*)$$

$L = E_{Kin} - E_{pot}$ -Lagrangian function

4. 2 Lagrangian equations

(Euler-Bernoulli beam)

Kinetic energy density $\mathcal{E}_{Kin} = \frac{1}{2} \mu \dot{u}^2$

Potential energy density $\mathcal{E}_{pot} = \frac{1}{2} EI (u'')^2$ } cf. 4.1

Lagrangian density $\mathcal{L} = \mathcal{E}_{Kin} - \mathcal{E}_{pot} = \mathcal{L}(\dot{u}, u'')$. (A)

Hamilton's principle:

$$\delta \int_{t_1}^{t_2} \int_0^l \mathcal{L} \, dX \, dt = 0. \qquad (*)$$

From (A) $$\delta \mathcal{L} = \frac{\delta \mathcal{L}}{\delta \dot{u}} \delta \dot{u} + \frac{\delta \mathcal{L}}{\delta u''} \delta u'' \qquad (B)$$

Putting (B) into (*) and integrating by parts we obtain

$$\underbrace{\int_0^l \frac{\partial \mathcal{L}}{\partial \dot{u}} \delta u \, dX \Big|_{t_1}^{t_2}}_{①} + \int_{t_1}^{t_2} \Bigg\{ \underbrace{\frac{\partial \mathcal{L}}{\partial u''} \delta u \Big|_0^l - \left(\frac{d}{dX} \frac{\partial \mathcal{L}}{\partial u''} \right) \delta u \Big|_0^l}_{②}$$

$$+ \int_0^l \underbrace{\left[\frac{d^2}{dX^2} \frac{\partial \mathcal{L}}{\partial u''} - \frac{d}{dt} \frac{\partial \mathcal{L}}{\partial \dot{u}} \right]}_{③} \delta u \, dX \Bigg\} dt = 0$$

① vanishes because $\delta u \big|_{t_1} = \delta u \big|_{t_2} = 0$ — (chosen)

② Natural (dynamical) boundary conditions:

$$\frac{\partial \mathcal{L}}{\partial u''}\, \delta u' \Big|_0^l - \left(\frac{d}{dX} \frac{\partial \mathcal{L}}{\partial u''} \right) \delta u \Big|_0^l = 0$$

(δu and $\delta u'$ have to satisfy the geometrical boundary conditions prescribed for u and u').

③ Lagrange equation (= Eulerian equation of the variational principle):

$$\frac{d}{dt} \frac{\partial \mathcal{L}}{\partial u} - \frac{d^2}{dX^2} \frac{\partial \mathcal{L}}{\partial u''} = 0 \hookrightarrow \text{equation of motion}$$

(δu arbitrary). Partial derivatives mean only differentiation with respect to explicit dependence, total derivatives mean differentiation with respect to implicit dependence too.

Lit. : H. Goldstein, Classical mechanics, Addison-Wesley, Cambridge, Mass., 1951

4. 3 Integral equation

Example : String

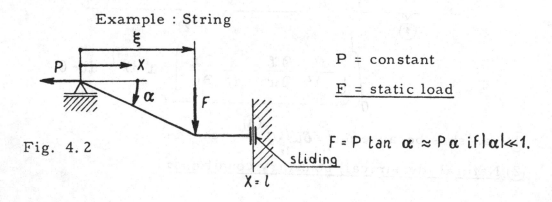

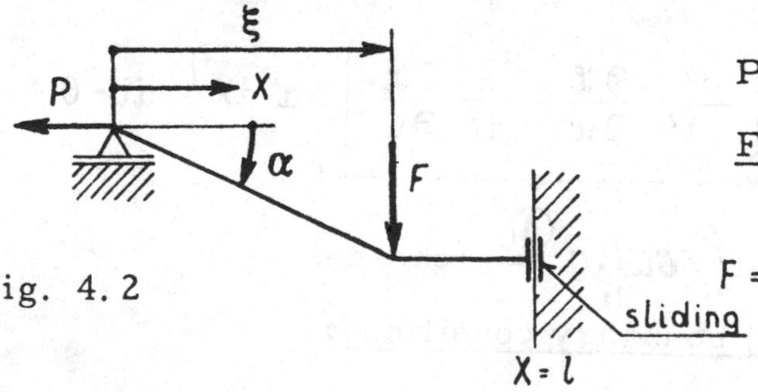

P = constant

F = static load

$F = P \tan \alpha \approx P\alpha$ if $|\alpha| \ll 1$.

Fig. 4. 2

$$\left.\begin{aligned} u &= X \cdot \frac{F}{P} \quad \text{for } 0 \leq X \leq \xi \\ u &= \xi \, \frac{F}{P} \quad \text{for } \xi \leq X \leq l \end{aligned}\right\} \; u = K(X, \xi) F \qquad (**)$$

Green's function

(Influence function)

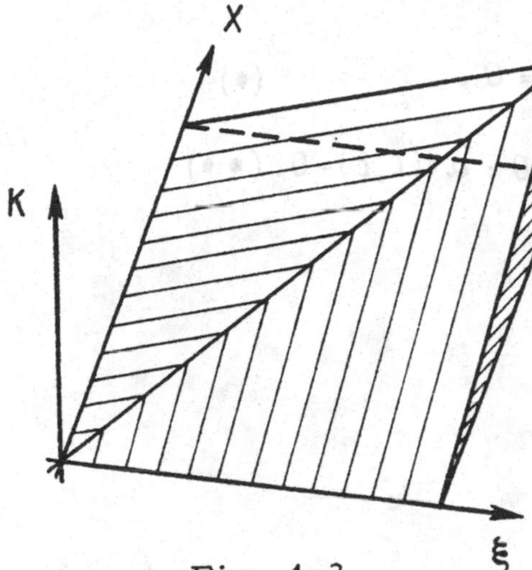

Fig. 4.3

Symmetry :

$K(X, \xi) = K(\xi, X)$

(Maxwell's reciprocity)

K is symmetric if the problem is self-adjoint; cf. sect. 5.4 and Collatz's book cited in section 5.5

Distributed load: $F(\xi) \hookrightarrow f(\xi)\, d\xi$

(**) is linear $\hookrightarrow$ superposition

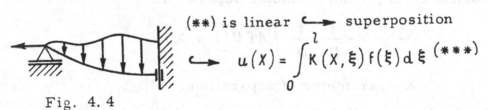

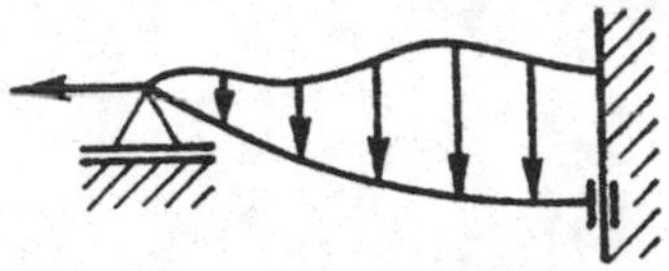

$$\hookrightarrow \quad u(X) = \int_0^l K(X, \xi) f(\xi)\, d\xi \qquad (***)$$

Fig. 4.4

Vibration $\hookrightarrow$ dynamic load : $f(\xi) = -\ddot{u}(\xi, t)\, \mu(\xi)$

$$u(X, t) = -\int_0^l K(X, \xi)\, \mu(\xi)\, \ddot{u}(\xi, t)\, d\xi .$$

5. Separation of variables; boundary value problem; eigenvalue problem

Example : Longitudinal vibrations of a rod

Equation of motion, cf. 2.125,

$$\mu \ddot{u} - \left[A E u' \right]' = 0 . \qquad (*)$$

Boundary conditions : $u(0,t) = 0 \; ; \; u'(l,t) = 0 .$ (∗∗)

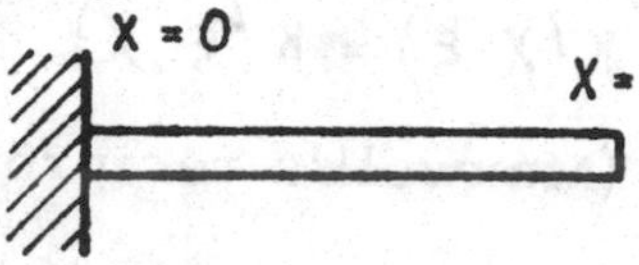

Fig. 5.1

5.1 Separation of the variables

$$u(X,t) = U(X)\, T(t) \qquad (***)$$

substituted in (∗) and variables separated :

$$-\frac{\ddot{T}}{T} = -\frac{1}{U}\frac{1}{\mu}\left[A E U' \right]' = \lambda$$

λ - parameter of separation .

We obtain two ordinary, second order differential equations :

$$\ddot{T} + \lambda T = 0 \qquad (A)$$

and

$$\left[A E U' \right]' + \lambda \mu = 0 \qquad (B)$$

From boundary conditions (∗∗) : $U(0) = 0 \, , \; U'(l) = 0 .$ (C)

5. 2 Boundary value problem - eigenvalue problem

Let $U_I(X,\lambda)$ and $U_{II}(X,\lambda)$ be two linearly independent solutions of (B). General solution of (B)

$$U(X,\lambda) = C_I U_I(X,\lambda) + C_{II} U_{II}(X,\lambda)$$

should satisfy the boundary conditions (C) :

$$\left.\begin{aligned} C_I U_I(0,\lambda) + C_{II} U_{II}(0,\lambda) = 0 \\ C_I U_I'(l,\lambda) + C_{II} U_{II}'(l,\lambda) = 0 \end{aligned}\right\} \text{(D)}$$

homogeneous system of equations for C_I, C_{II}

$$\Delta(\lambda) := \det \begin{Bmatrix} U_I(0,\lambda) & U_{II}(0,\lambda) \\ U_I'(l,\lambda) & U_{II}'(l,\lambda) \end{Bmatrix}$$

Characteristic equation :

$$\Delta(\lambda) = 0 \; ; \; roots \;\; \lambda_1, \lambda_2, \ldots$$

↳ eigenvalues,
characteristic values.

For $\lambda = \lambda_n \hookrightarrow C_I = (C_I)_n$, $C_{II} = (C_{II})_n \hookrightarrow$ solution $U = U_n$.
(eigenfunction)

(B) and (C) are satisfied by λ_n , U_n :

$$[A E U_n']' + \lambda_n \mu U_n = 0, \qquad \text{(E)}$$
$$U_n(0) = 0 \ , \ U_n'(l) = 0 .$$

((B) and (C) constitute a Sturm-Liouville eigenvalue problem).

In general :

$\Delta(\lambda) = 0$ - trascendental equation $\hookrightarrow$ infinitely many roots λ_n, $n = 1, 2, \ldots$, -denumerable

$\Delta(\lambda)$ - integral function of $\lambda \hookrightarrow$ no finite cluster points.

Lit. : R. Courant, D. Hilbert, Methods of mathematical physics, Vol. 1, Interscience Publ., N. Y. 1953

E. A. Coddington, N. Levinson, Theory of ordinary differential equations, Mc Graw-Hill, New York 1955

M. A. Neumark, Lineare Differentialoperatoren, Akademie-Verlag, Berlin 1963

5. 3 Example: Uniform rod

A, E, μ - constant

$$c^2 U'' + \lambda U = 0 \quad , \quad c^2 = AE/\mu \qquad \text{(B')}$$

General solution of (B')

$$U = C_I \sin \frac{\omega}{c} X + C_{II} \cos \frac{\omega}{c} X \, , \quad \omega = \sqrt{\lambda} \, .$$

Boundary conditions (C) :

$$C_I \cdot 0 \quad + \quad C_{II} \cdot 1 \ = 0$$

$$\frac{\omega}{c} C_I \cos \frac{\omega}{c} l + \frac{\omega}{c} C_{II} \sin \frac{\omega}{c} l = 0$$

Characteristic equation :

$$\Delta(\omega) = \frac{\omega}{c} \cos \frac{\omega}{c} l = 0$$

Roots : $\underbrace{\omega = 0}$ and $\quad \omega_n = \frac{c}{l} \pi \left(n - \frac{1}{2}\right) \, , \; n = 1, 2, \ldots$

$\hat{=}$ trivial sol. (negative n ↪ no new solutions)

$$(C_{II})_n = 0 \, , \; (C_I)_n =: C_n \ \text{arbitrary}$$

Solution of the eigenvalue problem

$$U_n(x) = C_n \underbrace{\sin \frac{\omega_n}{c} x}_{\text{eigenfunction, normal mode}}$$

Corresponding T_n from (A), p. 32,

$$T_n = a_n \sin \omega_n t + b_n \cos \omega_n t .$$

Special solutions of (*), (**), cf. 5.1,

$$u_n(x,t) = (A_n \sin \omega_n t + B_n \cos \omega_n t) \sin \frac{\omega_n}{c} x$$

$$A_n = a_n C_n \ , \quad B_n = b_n C_n \ \text{- arbitrary} \ ; \ n = 1,2 \ldots.$$

(A_n , B_n have to be calculated from the initial conditions, see sect. 5.5).

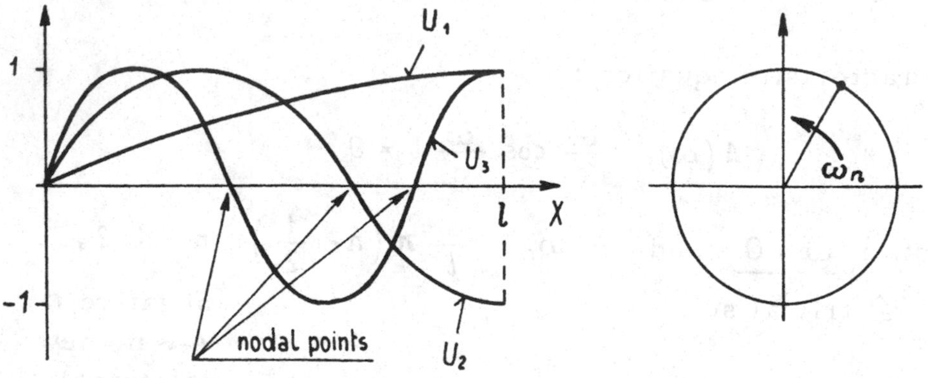

Fig. 5.2

5. 4 Adjoint and self-adjoint eigenvalue problem

5. 41 Inner product ; orthogonality, norm

Inner product:

$$(U,V) = \int_0^l U\bar{V}\,dX \,, \qquad \bar{V}\text{-complex conjug. of } V;$$

sometimes:

$$(U,V) = \int_0^l w(X)\,U\bar{V}\,dX\,; \quad w\text{-weight function};$$

for vectors :

$$(\underset{\sim}{U},\underset{\sim}{V}) = \int_0^l \underset{\sim}{U}^T\,\underset{\sim}{\bar{V}}\,dX\,; \qquad U^T\text{- transposed vector.}$$

(U, V) is the projection of U onto V, or vice versa. Two functions U, V are called orthogonal if

$$(U,V) = 0.$$

Norm:

$$\|U\| = (U,U)^{1/2} \geq 0\,,$$

U, V sufficiently smooth.

Lit. N.I. Achieser, I.M. Glaşmann, Theorie der liearen Operatoren im Hilbert-Raum, Akademie-Verlag, Berlin 1958.

Any text on functional analysis

5.42 Adjoint operator, adjoint boundary values, adjoint and self-adjoint eigenvalue problem

Equation (B) from 5.1 :

$$-\left[A E U'\right]' = \lambda \mu U.$$

Operator form

$$L U = \lambda \mu U.$$

L - linear differential operator,

$$(L U, V) = -\int_0^l \left[A E U'\right]' \bar{V} \, dX.$$

Integration by parts yields

$$(L U, V) = \underbrace{- A E U' \bar{V} \Big|_0^l + A E U \bar{V}' \Big|_0^l}_{\text{boundary terms (b.t.)}} \underbrace{- \int_0^l U \left[A E \bar{V}'\right]' dX}_{+(U, L^+ V)}$$

L^+ is the adjoint operator of L.

Here $L = L^+ \hookrightarrow$ L is a <u>self-adjoint operator</u>

Boundary terms :

$$\text{b.t.} = \underbrace{-AEU'(l)\bar{V}(l) - AEU(0)\bar{V}'0}_{\text{vanish because of (C), p. 32}} +$$

$$\underbrace{+AEU'(0)\bar{V}(0) + AEU(l)\bar{V}'(l)}_{\text{vanish for arbitrary } U'(0) \text{ and } U(l)}$$

<u>if</u>

$$(C^+)\left\{\begin{aligned} \bar{V}(0) &= 0 \hookrightarrow V(0)=0 \\ \bar{V}'(l) &= 0 \hookrightarrow V'(l)=0 \end{aligned}\right\}$$ boundary conditions adjoint to the boundary conditions (C) on page 32.

Here: "adjoint boundary conditions" = "original boundary conditions" $\longrightarrow$ <u>self-adjoint boundary conditions</u>.

<u>Adjoint eigenvalue problem :</u>

$$\begin{cases} \text{Adjoint differential equation } L^+U = \lambda\mu U \\ \text{adjoint boundary conditions } (C^+)\text{, see above.} \end{cases}$$

<u>Self-adjoint eigenvalue problem :</u>

$$\begin{cases} L^+ = L\,, \\ (C^+) = (C), \end{cases}$$

thus, see p. 38,

$$(LU, V) = (U, LV).$$

5.43 Eigenvalues of self-adjoint eigenvalue problems are real

Let U and λ be an eigenfunction and an eigenvalue, resp.,

$$\underbrace{(LU, U) - (U, LU)}_{=0} = \underbrace{(\lambda - \bar{\lambda})}_{=0!} \underbrace{\int_0^l \mu U \bar{U}\, dX}_{>0 \text{ because } \mu > 0 \text{ for all } X \text{ (positive definite)}}$$

$$\hookrightarrow \lambda = \bar{\lambda}$$

Since L is real $\hookrightarrow$ U may be assumed to be real.

5.44 Eigenfunctions belonging to different eigenvalues of self-adjoint eigenvalue problems are orthogonal

U_n, U_m - eigenfunctions

λ_n, λ_m - corresponding eigenvalues, $\lambda_n \neq \lambda_m$

$$\underbrace{(LU_n, U_m) - (U_n, LU_m)}_{=0} = \underbrace{(\lambda_n - \lambda_m)}_{\neq 0} \underbrace{\int_0^l \mu U_n U_m\, dX}_{=0!}$$

U_n, U_m are orthogonal with respect to the weight function $\mu(X)$; in general $\int_0^l U_n U_m\, dX \neq 0.$

5.45 Normalized eigenfunctions

Frequently it is convenient to normalize the eigen-functions in such a way that

$$\int_0^l \mu U_n^2 \, dX = 1$$

Then
$$\int_0^l \mu U_n U_m dX = \delta_{nm} = \begin{cases} 1 \text{ if } n = m \\ 0 \text{ if } n \neq m \end{cases}$$

For the example 5.3 we obtain because of $\mu \equiv 1$:

$$U_n = \sqrt{\frac{2}{l}} \sin \frac{\omega_n}{c} X .$$

5.5 Initial value problem

The rod shown in fig. 5.1 has initially, at $t = 0$

the displacement $u(X,0) = \varphi(X)$
the velocity $\dot{u}(X,0) = \psi(X)$ } cf. 3.111

(*), (**), cf. p. 32, has a solution of the form

$$u(X,t) = \sum_{n=1}^{\infty} (A_n \sin \omega_n t + B_n \cos \omega_n t) \, U_n(X) ; \qquad \text{(A)}$$

cf. p. 36; let the $U_n(x)$ be normalized, cf. above.
Substituting (A) into the initial conditions we obtain

$$\left.\begin{aligned} \varphi(X) &= \sum_{n=1}^{\infty} B_n U_n(X) \\ \psi(X) &= \sum_{n=1}^{\infty} A_n \omega_n U_n(X) \end{aligned}\right\} \qquad \text{(B)}$$

(B) represents two infinite systems of linear equations for A_n and B_n, $n = 1, 2, \ldots$.

Generalized Fourier coefficients

To solve (B) we multiply the equations by $\mu(X)\,U_m(X)$ and integrate over $0 \leq X \leq l$. Because of

$$\int_0^l \mu\, U_n U_m \, dX = \delta_{nm} \;,$$

we obtain

$$\left.\begin{aligned} B_m &= \int_0^l \varphi(X)\,\mu(X)\,U_m(X)\,dX \\ &\text{and} \\ A_m &= \frac{1}{\omega_m}\int_0^l \psi(X)\,\mu(X)\,U_m(X)\,dX . \end{aligned}\right\} \qquad \text{(C)}$$

$A_m \omega_m$ and B_m, given by (C), represent the generalized Fourier-coefficients of $\psi(X)$ and $\varphi(X)$, respectively, with respect to the system of functions $U_n(X)$.

The series (A), with the coefficients A_m, B_m, converges and satisfies the initial conditions if $\varphi(X)$ and $\psi(X)$ are sufficiently smooth and if the system $U_n(X)$ is complete ("expansion theorem").

General investigations :

Courant, Hilbert, Vol. 1
Coddington, Levinson,
Neumark
} cf. p. 34

Achieser, Glasmann, cf. p. 38

L. Collatz, Eigenwertemfgaben mit technischen Anwendungen, Akademische Verlagsgesellschaft, Leipzig 1949.

6. Forced vibrations

Example : Longitudinal vibrations of a rod.

6. 1 Excitation at one end (of a rod)

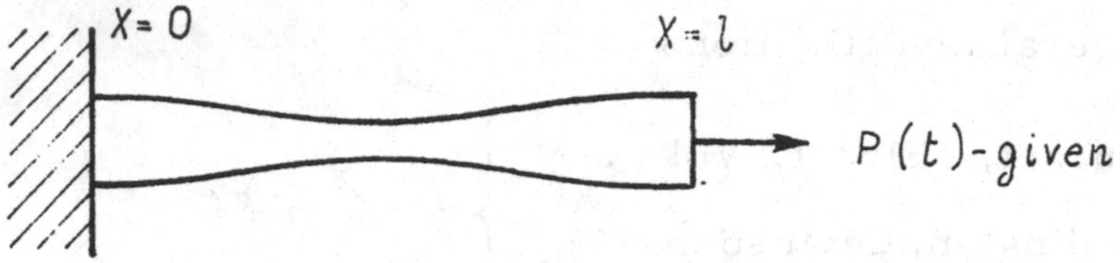

Fig. 6. 1

Equation of motion: $\mu \ddot{u} - \left[AEu'\right]' = 0$ (*)

Boundary conditions :

$$\left.\begin{aligned} &u(0,t) = 0 \\ &AE\,u'(l,t) - P(t) = 0 \quad \text{- nonhomogeneous} \end{aligned}\right\} (**)$$

Procedure : Superposition

$$u = u_p + u_n ,$$

where u_p is a particular solution which satisfies (*), (**), and u_n is the solution of the corresponding system with homogeneous boundary conditions, cf. (*), (**) on p. 32.

In general it is impossible to find u_p in closed form.

Trick : $u_p = u_I + u_{II}$.

u_I satisfies the (nonhomogeneous) boundary conditions (**) but not the differential equation (*),

u_{II} satisfies the differential equation (*), disturbed by u_I (see below), and the homogeneous boundary conditions

$$\left.\begin{aligned} u_{II}(0,t) &= 0, \\ u'_{II}(l,t) &= 0. \end{aligned}\right\} \quad (***)$$

We choose

$$u_I = \frac{P(t)}{E_l A_l} X \; ; \; E_l = E(l) \; ; \; A_l = A(l) \; ;$$

(**) is satisfied. Putting

$$u_p = \frac{P}{E_l A_l} X + u_{II} \qquad \text{into } (*) \qquad (A)$$

yields

$$\mu \ddot{u}_{II} - \left[A E u'_{II} \right]' = -\mu \frac{\ddot{P}}{E_l A_l} X + (AE)' \frac{P}{E_l A_l} \,. \qquad (B)$$

This is a nonhomogeneous differential equation of the form

$$\mu \ddot{u} - \left[A E u' \right]' = p(X,t)$$

which corresponds to a system with distributed excitation, cf. section 6. 2.

6. 2 Distributed excitation

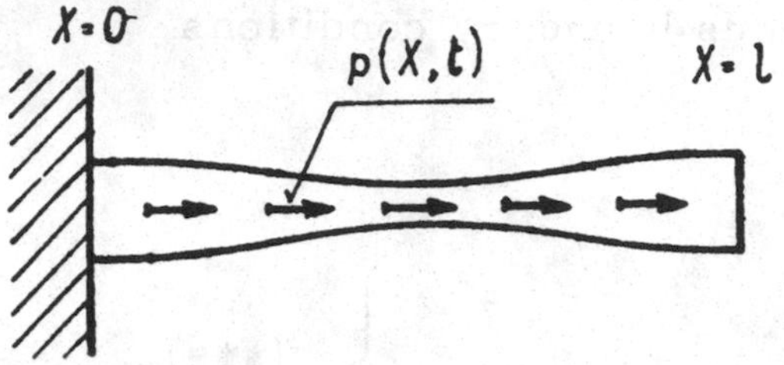

Fig. 6. 2

Equation of motion :

$$\mu \ddot{u} - \left[A E u'\right]' = p(X,t). \qquad (*)$$

Homogeneous boundary conditions :

$$\left.\begin{aligned} u(0,t) &= 0 \\ u'(l,t) &= 0 \end{aligned}\right\} \qquad (**)$$

6. 21 Normal coordinates

A solution of (*), (**) is sought in the form

$$u(X,t) = \sum_{n=1}^{\infty} a_n(t)\, U_n(X) \qquad \text{(A)}$$

U_n - eigenfunctions of the homogeneous problem corresponding to (*), (**), cf. 5. 2.

a_n - normal coordinates - to be sought

Putting (A) into (*) we obtain

$$\sum_{n=1}^{\infty} \left\{ \mu \ddot{a}_n U_n - a_n \left[A E U_n' \right]' \right\} = p(X,t).$$

Application of the relation (E) from p. 34,

$$\left[A E U_n' \right]' = - \lambda_n \mu U_n ,$$

yields

$$\sum_{n=1}^{\infty} \left\{ \ddot{a}_n + \lambda_n a_n \right\} \mu U_n = p(X,t) \; ; \; \lambda_n = \omega_n^2 .$$

Fourier expansion, cf. 5. 45, 5. 5, leads to

$$\ddot{a}_n + \omega_n^2 a_n = b_n \; , \; n = 1, 2, \ldots , \qquad \text{(B)}$$

where

$$b_n(t) = \int_0^l p(X,t)\, U_n(X)\, dX . \qquad \text{(C)}$$

(B) is an decoupled system of infinitely many ordinary non-homogeneous differential equations with constant coefficients.

Solutions of (B) :

$$a_n(t) = \underbrace{A_n \sin \omega_n t + B_n \cos \omega_n t}_{a_{n_h}} + \underbrace{\frac{1}{\omega_n}\int_0^t \sin \omega_n (t-\tau)\, b_n(\tau)\, d\tau}_{a_{n_p}} \tag{D}$$

A_n, B_n from initial conditions, cf. sect. 5.5.

6.22 Resonance

$b_n(t)$ periodic, cf. (C); for instance

$$b_n = \beta_n \sin \Omega t.$$

From (D) :

$$a_{n_p} = \frac{\beta_n}{\omega_n^2 - \Omega^2} \sin \Omega t.$$

Large amplitude — resonance — for $\Omega \to \omega_n$ if $\beta_n \neq 0$. Dangerous region : $|\omega_n - \Omega|$ small.

Technically important in rotating and reciprocating machines.

7. Methods to calculate eigenvalues of conservative systems

Separation of variables, cf. sect. 5.1, transforms the equations of motion of section 2 into

$$L\,U = \lambda \mu U, \qquad (*)$$

where L and μ may be matrices and U may be a vector, e.g., in sect. 2.23.

The –homogeneous– boundary conditions have the form

$$\ell\,U = 0 \qquad (**)$$

7.1 Analytical method

An analytical solution of the differential equation (*) and a subsequent analytical or numerical solution of the characteristic equation are only possible for some very special systems. Realistic models have to be investigated by approximate methods.

7.2 Rayleigh's quotient

7.21 Formal introduction

Multiplication of (*) by U and integration over X yield :

$$\int_0^l (LU)\, U\, dX = \lambda \int_0^l U \mu U dX$$

This is a scalar equation;

$$\int_0^l U \mu U\, dX > 0 \quad if \quad U \not\equiv 0 \text{ and } \mu > 0$$

(positive definite).

We obtain Rayleigh's quotient

$$\lambda = \frac{\int_0^l (LU) U\, dX}{\int_0^l U \mu U\, dX} . \qquad \text{(A)}$$

If we substitute the eigenfunction U_n for U, we obtain by(A) the eigenvalue $\lambda = \lambda_n$. If we choose any function $U(X)$ which satisfies the boundary conditions (**),we get from (A) an approximate value, λ_1^* , for the first eigenvalue, λ_1 ; λ_1^* is always greater than or equal to λ_1 , $\lambda_1^* \geqslant \lambda_1$.
If we put a function U into (A),which satisfies the boundary conditions and is orthogonal to the first $(n-1)$ eigenfunctions,

$$\int_0^l U_k \mu U\, dX = 0 \ for \quad k = 1, \dots, n-1 ,$$

we obtain from (A) an approximation, λ_n^* , for the eigenvalue λ_n ; $\lambda_n^* \geq \lambda_n$.

Lit. : Collatz, Eigenwertaufgaben, cf. p.43.
J. P. Den Hartog, Mechanical vibrations, 4. ed., McGraw-Hill, New York 1956
R. Zurmühl, Praktische Mathematik, Springer, Berlin 1965

7.22 Introduction of Rayleigh's quotient by energy considerations

Conservative system :

$$E_{Kin} + E_{pot} = E_{tot} = \text{constant};$$

cf. 4.1.

Assuming an "in phase" sinusoidal motion of the whole system we have

$$E_{pot/max} = E_{tot} \quad \text{at maximal deflection, (zero velocity)}$$

$$E_{Kin/max} = E_{tot} \quad \text{at maximal velocity, (zero deflection),}$$

or $$E_{pot/max} = E_{Kin/max} .$$

Example. Euler-Bernoulli beam, cf. 2.22

$$E_{Kin} = \frac{1}{2}\int_0^l \dot{u}^2 \mu \, dX = \frac{1}{2}\omega^2 \cos^2 \omega t \int_0^l U \mu U \, dX ,$$

$$E_{Kin/max} = \frac{1}{2}\underbrace{\lambda}_{\omega^2} \int_0^l U \mu U \, dX ;$$

$$E_{pot} = \frac{1}{2}\int_0^l EI (u'')^2 \, dX = \frac{1}{2}\sin^2 \omega t \int_0^l U'' EI U'' \, dX$$

$$E_{pot/max} = \frac{1}{2}\int_0^l U'' EI U'' \, dX$$

Equating these expressions for $E_{Kin/max}$ and $E_{pot/max}$, cf. above, we obtain

$$\lambda = \frac{\int_0^l U'' EI U'' \, dX}{\int_0^l U \mu U \, dX} . \qquad \text{(B)}$$

The denominator of this quotient has the form of the denominator in section 7.21. The numerator differs from the numerator in 7. 21 only by some partial integrations (if the boundary conditions are self-adjoint). So there are only formal differences between (A) and (B).

In section 5. 43 we showed that self-adjoint eigenvalue problems have real eigenvalues. Looking at the numerator and the denominator of (B), we see, because of their mechanical meaning, that both of them must be positive definite (they are ≥ 0 for any choice of real functions U); thus, λ must be positive (ω real, $\omega^2 = \lambda$); cf. sect. 8.

7. 23 Rayleigh-Ritz method

Because of $\lambda_1^* \geq \lambda_1$ for all admissible functions U, we may conclude

$$\lambda_1 = \min_{\text{all admiss } U} \frac{\int_0^l (LU)\, U \, dX}{\int_0^l U \mu U \, dX} \qquad \text{(C).}$$

Ritz : Introduce a function $U(X, \alpha_1, \ldots, \alpha_M)$ which depends on some parameters $\alpha_1, \ldots, \alpha_M$ and satisfies the boundary conditions for arbitrary α_m, $m = 1, .., M$. By (A) or (B) we obtain $\lambda_1^* = \lambda_1^*(\alpha_1, \ldots, \alpha_M) \geq \lambda_1$.

The best approximation for λ_1 we find from

$$\min_{\alpha_1,\dots,\alpha_M} \lambda_1^*(\alpha_1,\dots,\alpha_M).$$

This is an ordinary minimum problem. Further investigations and simplifications see in Zurmühl, Praktische Mathematik, cf. p. 51.

Error estimates : H. Schellhaas, Ein Verfahren zur Berechnung von Eigenwertschranken mit Anwendungen auf das Beulen von Rechteckplatten, Ing.-Arch. 37(1968) 243-250

S. G. Michlin, Variationsmethoden der mathematischen Physik, Akademie Verlag, Berlin 1962

7. 3 Iteration process

Procedure : Choose zero approximation $U^{(0)}(X)$ and calculate $U^{(1)}(X)$ from

$$\left.\begin{aligned} L U^{(m+1)} &= \mu U^{(m)}, \quad m = 0, 1, \dots \\ \ell U^{(m+1)} &= 0, \end{aligned}\right\} \quad \text{(D)}$$

cf. (*) and (**) p. 49.

If Green's function $K(X,\xi)$ is known, cf. sect. 4.3, the procedure (D) can be replaced by

$$U^{(m+1)}(X) = \int_0^l K(X,\xi)\,\mu(\xi)\,U^{(m)}(\xi)\,d\xi \;, \; m = 0,1,\dots$$

which is deduced from the integral equation corresponding to (*), (**).

The procedure converges to the lowest eigenfunction, U_1. The respective eigenvalue can be expressed by Rayleigh's quotient.

Higher order eigenfunctions, $U_n, n > 1$, and eigenvalues, λ_n, can be calculated if $U^{(o)}$ and, subsequently, because of the unavoidable inaccuracies in the calculations, the $U^{(m)}$ are orthogonalized with respect to the lower eigenfunctions.

A semigraphical algorithm was introduced by Stodala, see Den Hartog, p. 51. Numerical procedures are given in Zurmühl's book, cf p. 51.

Variations of this procedure are possible(e.g., Grammel's procedure, cf. Zurmühl).

7.4 Variational methods

7.41 Ritz's method

In section 4.1 we replaced the equations of motion - for conservative systems - by Hamilton's principle,

$$\int_{t_1}^{t_2} \left(E_{Kin} - E_{pot} \right) dt \qquad = \text{extremum},$$

a variational principle.

Choosing the example "Euler-Bernoulli beam" from section 7.22 we have

$$\int_{t_1}^{t_2} \frac{1}{2} \int_0^l \left\{ \mu \dot{u}^2 - EI\,(u'')^2 \right\} dX\, dt = \text{extremum}.$$

Assuming $u = U(X) \sin \omega t,\ t_1 = 0,\ t_2 = 2\pi/\omega$

we obtain $\int_0^l \left\{ \lambda \mu U^2 - EI\,(U'')^2 \right\} dX = \text{extremum}, \quad \lambda = \omega^2 .$

We solve this problem by Ritz's direct method:

$$U(X) = \sum_{m=1}^{M} \alpha_m U_m^*(X), \qquad \text{(E)}$$

the U_m^* have to satisfy only the geometric boundary conditions and are linearly independent.

From $\delta \int_0^l \left\{ \mu \lambda U^2 - EI\,(U'')^2 \right\} dX = 0$

and $\delta U = \sum_{m=1}^{M} \delta \alpha_m U_m^* , \quad \delta \alpha_m$ - *arbitrary*, we obtain a set of linear homogeneous equations for α_m :

$$\left\{ \lambda \begin{pmatrix} A_{11} & A_{12} & \cdots & A_{1M} \\ A_{21} & & & \vdots \\ \vdots & & & \vdots \\ A_{M1} & \cdots & & A_{MM} \end{pmatrix} - \begin{pmatrix} B_{11} & B_{12} & \cdots & B_{1M} \\ B_{21} & & & \vdots \\ \vdots & & & \vdots \\ B_{M1} & \cdots & & B_{MM} \end{pmatrix} \right\} \begin{pmatrix} \alpha_1 \\ \alpha_2 \\ \vdots \\ \alpha_M \end{pmatrix} = 0, \quad (F)$$

where

$$A_{jk} = \int_0^l \mu U_j^* U_k^* \, dX = A_{kj} ,$$

$$B_{jk} = \int_0^l EI \, U_j'' \, U_k'' \, dX = B_{kj} .$$

For nontrivial solutions $(\alpha_m \neq 0 \, , \; m = 1, \ldots, M)$ of equation (F),

$$\Delta^* (\lambda) = \det \{ \ldots \} = 0 \qquad (G)$$

must vanish. (G) is an algebric approximation of the characteristic equation. Its roots, $\lambda_1^* , \ldots , \lambda_M^*$, are approximations for the eigenvalues $\lambda_1 , \ldots , \lambda_M$. (The approximations for the lower eigenvalues are better than those for the higher ones). The solution vectors $\begin{pmatrix} \alpha_1 \\ \vdots \\ \alpha_M \end{pmatrix}$ of (F), introduced into (E), yield approximate eigenfunctions.

The same set of equations, (F), (G), can be obtained by the Rayleigh-Ritz method, cf. 7.23, if (E) is introduced into (C), p. 53 ; cf. Zurmühl, Praktische Mathematik.

7.42 Galerkin's method

Galerkin's method starts from the differential equation (*), p. 49 , in the form

$$LU - \lambda\mu U = 0. \qquad \text{(H).}$$

If U, λ is a solution of (H) which satisfies the boundary conditions (**), then the projection, cf. 5.41, of (H) onto an arbitrary function $V(X)$ must vanish :

$$(LU - \lambda\mu U, V) = 0 \qquad \text{(I)}$$

(For the Euler-Bernoulli beam : $LU = EIU^{IV}$)

If we introduce an approximation of the form (E),

$$U(X) = \sum_{m=1}^{M} \alpha_m U_m^*(X) \qquad \text{(E')}$$

where the U_m^* have to satisfy all boundary conditions(**), and if we choose $V = U_m^*(X)$, $m = 1, \ldots, M$, we obtain the equations (F), p. 57, again (with the same coefficients A_{jk}, B_{jk} if the $U_m^*(X)$ in (E) and (E') are the same). Thus, the equations (F) are sometimes called Ritz-Galerkin equations. Subsequent calculations cf. 7.41.

7. 43 Differences between Ritz's and Galerkin's method

	Ritz	Galerkin
Basis	Variational principle	Differential equation (principle of virtual work)
Approximate solution cf. (E), (E'), has to satisfy	the geometrical boundary conditions	all boundary conditions (can be weakened)
δU (Ritz) V (Galerkin)	must be derived from the approximate solution	can be chosen arbitrarily (have to satisfy only the geometrical boundary conditions if the principle of virtual work is used)
Applicable	to problems which can be expressed as variational problems	to problems which are governed by differential equations or the principle of virtual work

Lit. : L. W. Kantorovitsch, W. I. Krylow,

Näherungsmethoden der höheren Analysis, VEB

Deutscher Verlag der Wissenschaften,

Berlin 1956

7. 5 Transfer matrices (Example: Euler - Bernoulli beam)

7. 51 Basic idea

Example : Euler-Bernoulli beam

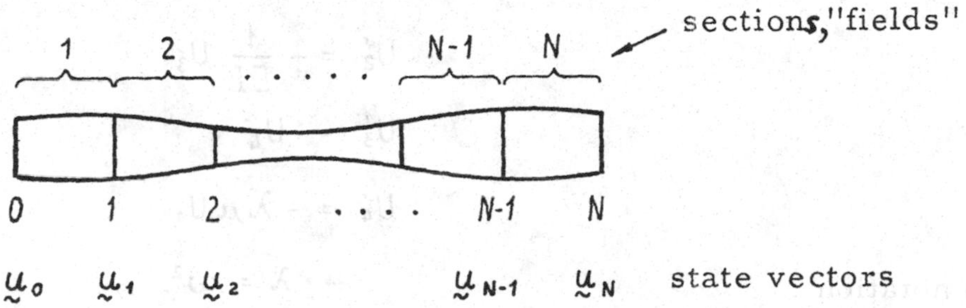

Fig. 7. 1

Linear relation

$$\underset{\sim}{u}_n = \underset{\sim}{F}_n^* \, \underset{\sim}{u}_{n-1} \quad , \quad \underset{\sim}{F}_n^* \text{ operator}$$

7. 52 Basic equations

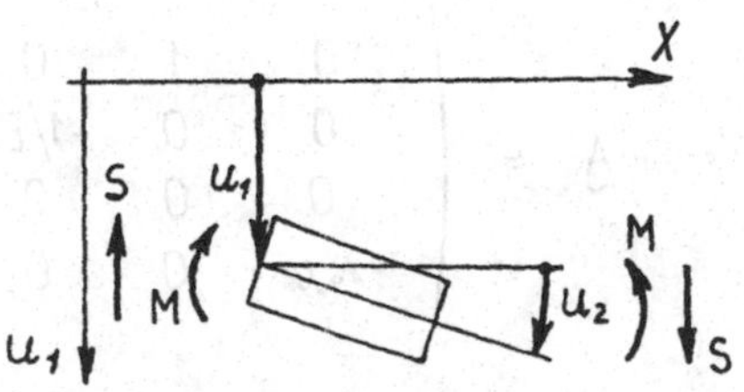

Fig. 7. 2

u_1 - lateral displacement

u_2 - angular displacement

$u_3 = M$ - bending moment

$u_4 = S$ - shear force

EI - bending stiffness

μ - mass per unit length

$$u_1' = u_2$$

$$u_2' = -\frac{M}{EI} = -\frac{u_3}{EI}$$

$$u_3' = S = u_4$$

$$u_4' = \mu \ddot{u}_1$$

Assumption :

$$u_j = U_j \sin \omega t$$

$$U_1' = U_2$$

$$U_2' = -\frac{1}{EI} U_3$$

$$U_3' = U_4$$

$$U_4' = -\lambda \mu U_1$$

$$\lambda = \omega^2.$$

Matrix notation :

$$\underset{\sim}{U}' = \underset{\sim}{A}(\lambda, X)\, \underset{\sim}{U},$$

where

$$\underset{\sim}{U} = \{U_1, U_2, U_3, U_4\} \quad - \quad \text{state vector}$$

$$\underset{\sim}{A} = \begin{pmatrix} 0 & 1 & 0 & 0 \\ 0 & 0 & -1/EI & 0 \\ 0 & 0 & 0 & 1 \\ -\lambda\mu & 0 & 0 & 0 \end{pmatrix}$$

Field n with EI, μ - constant, initial point $X = 0$, term

inal point $X = l_n$:

$$\underset{\sim}{U}_n := \underset{\sim}{U}_n(l) = e^{\underset{\sim}{A}_n l_n} \cdot \underset{\sim}{U}_n(0) = e^{\underset{\sim}{A}_n l_n}\, \underset{\sim}{U}_{n-1} \, ,$$

or $\qquad \underset{\sim}{U}_n = \underset{\sim}{F}_n \underset{\sim}{U}_{n-1} \qquad \underset{\sim}{F}_n$ - field matr

$$\underset{\sim}{F}_n = e^{\underset{\sim}{A}_n l_n} = \underset{\sim}{F}_n(\lambda) \qquad \text{(K)}$$

Successive applications of (K) yield

$$\underset{\sim}{U}_N = \underset{\sim}{F}_N \underset{\sim}{U}_{N-1} = \underset{\sim}{F}_N \underset{\sim}{F}_{N-1} \underset{\sim}{U}_{N-2} = \underset{\sim}{F}_N \underset{\sim}{F}_{N-1} \cdots \underset{\sim}{F}_2 \underset{\sim}{F}_1 \underset{\sim}{U}_0 \qquad \text{(L)}$$

7.53 Boundary conditions

Let us choose the following boundary conditions (cf. Fig. 7.3)

$$\left.\begin{array}{l} \underset{\sim}{U}_0 : (U_1)_0 = 0, (U_2)_0 = 0 \hookrightarrow \underbrace{\begin{pmatrix} 1 & 0 & 0 & 0 \\ 0 & 1 & 0 & 0 \end{pmatrix}}_{\underset{\sim}{B}_0} \underset{\sim}{U}_0 = 0 \qquad \underset{\sim}{B}_0 \underset{\sim}{U}_0 = 0 \\ \underset{\sim}{U}_N : (U_3)_0 = 0, (U_4)_0 = 0 \hookrightarrow \begin{pmatrix} 0 & 0 & 1 & 0 \\ 0 & 0 & 0 & 1 \end{pmatrix} \underset{\sim}{U}_N = 0 \qquad \underset{\sim}{B}_N \underset{\sim}{U}_N = 0 \end{array}\right\} \text{(M)}$$

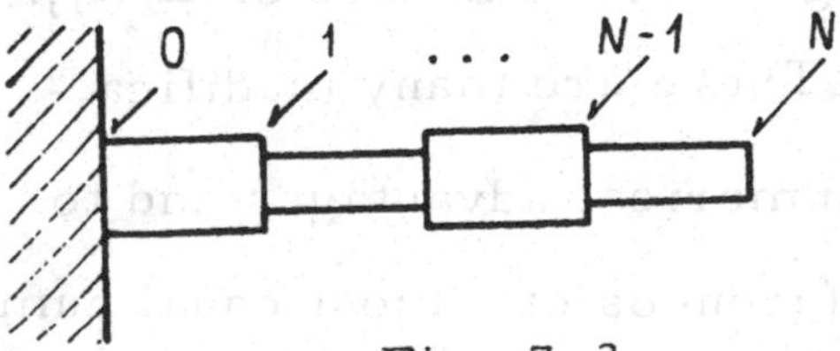

Fig. 7.3

7.54 Characteristic equation

Introducing (L) into the boundary conditions (M) we obtain

$$\underbrace{\begin{pmatrix} \underset{\sim}{B}_0 \\ \underset{\sim}{B}_N \underset{\sim}{F}_N \underset{\sim}{F}_{N-1} \cdots \underset{\sim}{F}_2 \underset{\sim}{F}_1 \end{pmatrix}}_{\text{hypermatrix } \underset{\sim}{M}} \underset{\sim}{U}_0 = 0 \qquad \text{(N)}$$

r

$$\underset{\sim}{M}(\lambda)\, \underset{\sim}{U}_0 = 0. \qquad \text{(N')}$$

(N) is a linear homogeneous system of equations for $\underset{\sim}{U}_0$ We obtain the characteristic equation

$$\Delta(\lambda) = \det \underset{\sim}{M}(\lambda) = 0 \qquad \text{(P)}$$

Equation (P) can be solved numerically on a digital computer. A value λ^* is chosen, the field matrices $\underset{\sim}{F}_n(\lambda^*)$ are calculated, cf. equ. (K). Finally $\Delta(\lambda^*)$ is calculated by equations (N), (P) and plotted, cf. Fig. 7.4. The zeroes of $\Delta(\lambda)$, λ_K, are eigenvalues of the system. There are many modifications of this procedure to gain numerical advantages and to avoid numerical difficulties (differences of almost equal numbers).

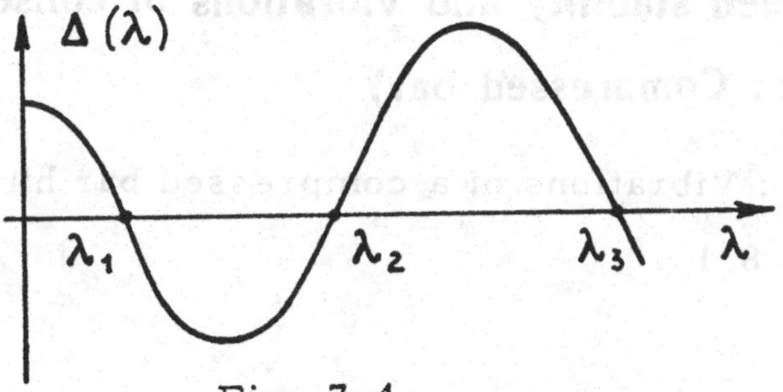

Fig. 7.4

General considerations and extensions :

E. C. Pestel, F. A. Leckie, Matrix methods in elastomechanics, McGraw-Hill, New York 1963

W. C. Hurty, M. F. Rubinstein, Dynamics of structures, Prentice - Hall, Englewood Cliffs, N. J., 1964

Numerical difficulties : K. Marguerre, R. Uhrig, ZAMM 44 (1964), pp. 1-21 and 349-360

8. Relations between stability and vibrations of conservative systems (Example: Compressed bar)

Example : Vibrations of a compressed bar hinged at its ends, cf. Fig. 8.1

8.1 Equations of motion; boundary conditions

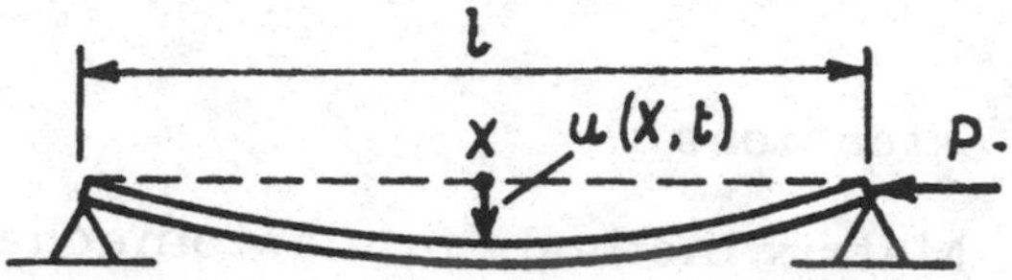

Fig. 8.1

$$\left.\begin{matrix} EI \\ \mu \end{matrix}\right\} \text{- constant}$$

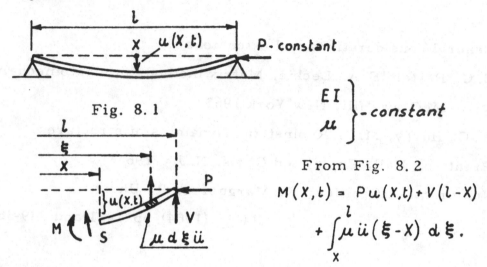

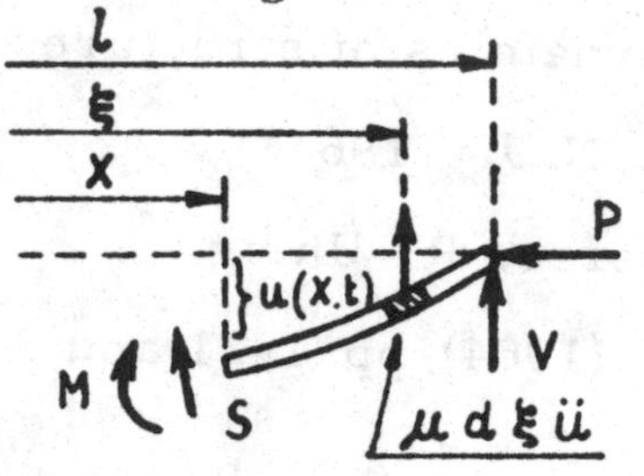

Fig. 8.2

From Fig. 8.2

$$M(x,t) = P u(x,t) + V(l-x) + \int_x^l \mu \ddot{u}(\xi - x)\, d\xi .$$

Moment curvature relation :

$$M = -EI u''$$

Elimination of M yields the equation of motion :

$$EI u^{IV} + P u'' + \mu \ddot{u} = 0 . \qquad (*)$$

Boundary conditions:

$$\left.\begin{aligned} & u(0,t) = u(l,t) = 0 \\ & M(0,t) = M(l,t) = 0 \hookrightarrow u''(0,t) = u''(l,t) = 0 \end{aligned}\right\} \quad (**)$$

Trivial solution of (*), (**) : $u(X,t) \equiv 0$. (***)

8. 2 Eigenfunctions and natural frequencies

Separation of the variables, $u(X,t) = U(X)e^{\gamma t}$,

leads to the eigenvalue problem

$$EI\,U^{IV} + PU'' - \lambda\mu U = 0, \quad (\lambda = -\gamma^2 = \omega^2,\ \omega = i\gamma), \quad (C)$$

$$\left.\begin{aligned} U(0) &= U(l) = 0, \\ U''(0) &= U''(l) = 0, \end{aligned}\right\} \quad (D)$$

(self-adjoint eigenvalue problem).

Solutions of (C), (D) :

Eigenfunctions : $U_n = \sin \frac{n\pi}{l} X$ (not normalized)

Eigenvalues : $\lambda_n = \omega_n^2 = -\gamma_n^2 = \frac{1}{\mu}\frac{EI n^4 \pi^4}{l^4}\left[1 - \frac{Pl^2}{EI n^2 \pi^2}\right]$

8. 3 Discussion of the results

Distribution of the eigenvalues $\gamma_n\ (= -i\omega_n)$ see Fig. 8. 3/4 .

The eigenvalues move as indicated by the arrows if Pl^2/EI is increased.

For $Pl^2/EI > 1$ $\gamma_1 \hookrightarrow$ real, positive, the first normal mode, $U_1(X)$, increases exponentially $\hookrightarrow$ the trivial solution becomes unstable.

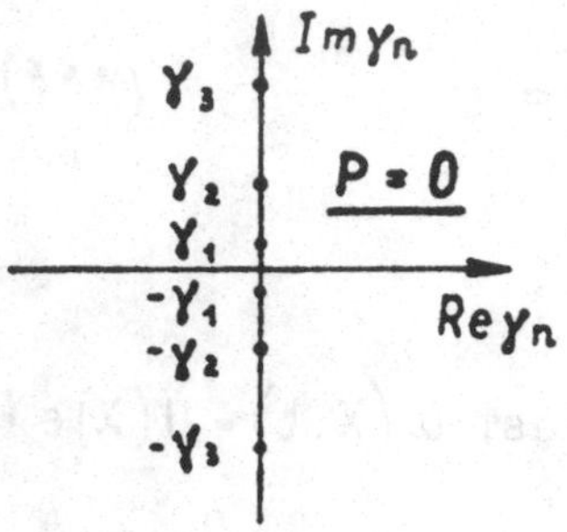

Fig. 8.3

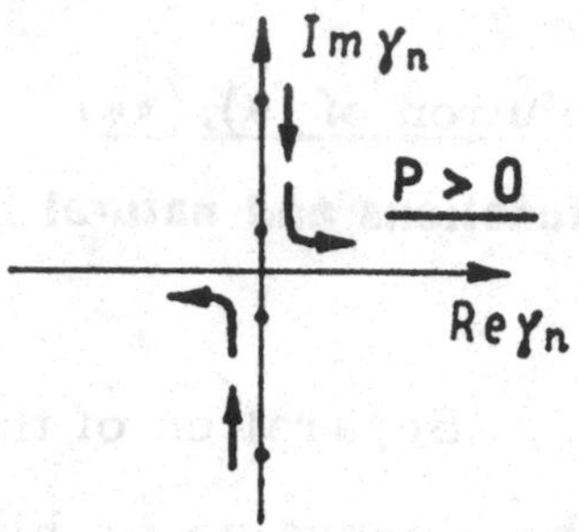

Fig. 8.4

Border line between stable and unstable trivial solutions: $\gamma_1 = 0$; the eigenvalue problem reduces to:

$$EI\,U^{IV} + PU'' = 0 \qquad (C')$$

$$\left.\begin{aligned} U(0) &= U(l) = 0 \\ U''(0) &= U''(l) = 0 \end{aligned}\right\} \qquad (D')$$

P - parameter (eigenvalue).

Solutions U_n , P_n of (C'), (D') represent nontrivial static deflections of the rod.

Static stability criterion : The trivial solution (state of equilibrium) of the system becomes unstable for $P > P_{crit}$, where P_{crit} is the lowest value of the parameter (load) for which exists a nontrivial static deflection in the vicinity of the trivial solution (P_{crit} is the lowest eigenvalue of (C'), (D')).

9. Equations of motion for (nonlinear) nonconservative systems

9. 1 Differential equations; boundary conditions

Differential equations and boundary conditions can be established as in section 2. 1. Nonlinear terms arise from geometrical nonlinearities and from nonlinear constitutional equations.

9. 2 Principle of virtual work

$$\delta W_I + \delta W_W = 0 \qquad \text{(A)}$$

δW_I - virtual work of the intertia forces (D'Alembert's forces)

δW_W - virtual work of the "working" forces ("applied" forces in Goldstein's book cited in sect. 4. 2).

The (infinitesimally small) virtual displacements have to satisfy the geometrical constraints (boundary conditions).

Example : Extensible (linear) viscoelastic string (plane vibrations)

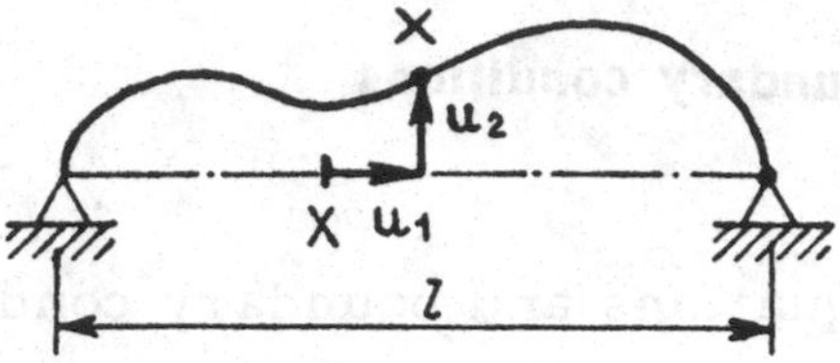

Fig. 9. 1

l - length of the unextended string

$$\varepsilon = \sqrt{(1+u_1')^2 + (u_2')^2} - 1 \tag{B}$$

$$F = A(E_1\varepsilon + E_2\dot{\varepsilon}) \quad \text{- force} \tag{C}$$

A - cross-sectional area

μ - mass per unit length

E_1, E_2 - constants

Virtual work of the intertia forces :

$$\delta W_I = -\int_0^l \mu\,(\ddot{u}_1\,\delta u_1 + \ddot{u}_2\,\delta u_2)\,dX \tag{D}$$

Virtual work of the working forces :

$$\delta W_w = -\int_0^l F\,\delta\varepsilon\,dX\ , \tag{E}$$

where F is taken from equation (C), and

$$\delta\varepsilon = \frac{(1+u_1')\,\delta u_1' + u_2\,\delta u_2'}{\sqrt{(1+u_1')^2 + (u_2')^2}}\ , \quad \text{cf. equation (B);} \tag{F}$$

$\delta u_i' = \frac{\partial}{\partial X}(\delta u_i)$, $i = 1,2$ - partial derivatives of the virtual displacements δu_i

The virtual displacements $\delta u_i = \delta u_i(X)$ are arbitrary functions which need not depend on the time, t; especially, the δu_i are not related to the actual displacements, $u_i(X,t)$; but both the actual displacements, u_i, and the virtual displacements, δu_i, are subjected to the geometrical boundary conditions (cf. Fig. 9.1) :

$$\left.\begin{array}{l} u_1(0,t) = u_1(l,t) = u_2(0,t) = u_2(l,t) = 0 \\ \delta u_1(0) = \delta u_1(l) = \delta u_2(0) = \delta u_2(l) = 0 \end{array}\right\} \text{(G)}$$

Instead of the double symbol $\delta u = \delta u(X)$ we could write, for instance, $v(X)$. But the δ in front of the u, etc. indicates and reminds that the virtual quantities have the character of differentials; the relations between the virtual quantities are relations between differentials, cf. the relation (F) between $\delta\varepsilon$, δu_1, δu_2 above.

The principle of virtual work, (A), can be used as basic equation for Galerkin's method, cf. sect. 7.42. The equation (A) is always linear with respect to the virtual displacements and can be interpreted as a projection which has to vanish for arbitrary, sufficiently smooth, functions δu (=virtual displacements). If (A) is used for Galerkin's method, both the approximate solution and the "projection" functions (in section 7. 42 denoted by U and V, respectively) have to satisfy only the geometrical conditions (cf. table in section 7. 43).

A general expression for the virtual work of the stresses in a (classical) continuous body is given in

A. E. Green, W. Zerna, Theoretical elasticity, Clarendon Press, Oxford 1963; (formula 2. 6. 7).

9. 3 The general form of Hamilton's principle

We obtain Hamilton's principle from the principle of virtual work by introducing time-dependent virtual displacements, e. g., $\delta u(X,t)$, and integrating (A) over a time interval $t_1 \leq t \leq t_2$:

$$\int_{t_1}^{t_2} (\delta W_I + \delta W_W)\, dt = 0 \tag{H}$$

By partial integration with respect to the time and suitable assumptions about the virtual displacements at

t_1 and t_2 it is always possible to express the first term of (H) by means of the kinetic energy :

$$\int_{t_1}^{t_2} \delta W_I \, dt = \int_{t_1}^{t_2} \delta E_{Kin} \, dt = \delta \int_{t_1}^{t_2} E_{Kin} \, dt , \qquad (I)$$

cf. section 4. 1.

If the system is conservative , δW_W too can be written as a variation of the potential energy, E_{pot} ,

$$\delta W_W = - \delta E_{pot} , \qquad (J)$$

and we obtain immediately Hamilton's principle in one of its usual forms which are given in section 4. 1.

But the usual form of Hamilton's principle can be valid for special nonconservative systems too : If δW_W can be considered as the variation of a time-dependent force function $U_F(u_1,...,t)$,

$$\delta W_n = \delta U_F ,$$

then Hamilton's principle can be written in one of the forms

$$\delta \int_{t_1}^{t_2} \left(E_{Kin} + U_F \right) dt = 0 , \qquad (K)$$

or

$$\int_{t_1}^{t_2} (E_{kin} + U_F)\, dt = \text{extremum}. \qquad \text{(K')}$$

Such a system is not conservative but (K) and (K') hold.

Lanczos calls such systems "monogenic" (the forces are generated by a single function); cf. C. Lanczos, The Variational Principles of Mechanics, 3. ed., University of Toronto Press, Toronto 1966.

Example for a monogenic system : The beam loaded by a pulsating force, cf. Fig. 9. 2

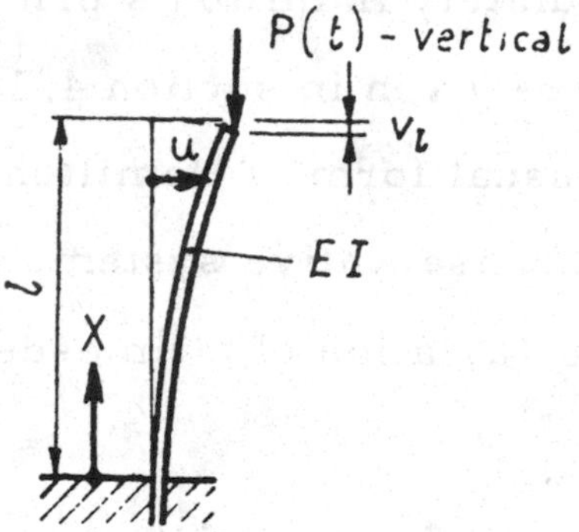

Fig. 9. 2

E I - bending stiffness (elastic beam), no longitudinal deformation; small deflections $u(X, t)$.

Elastic potential energy : $E_{pot} = \frac{1}{2} \int_0^l EI\, (u'')^2 dX$

Let v_l denote the vertical displacement at $X = l$

Then

$$U_F = P v_l - E_{pot}$$

(cf. the signs in (K), above, and in (*), section 4.1).

For small $u'(X,t)$ we have

$$v_l = \frac{1}{2}\int_0^l (u')^2\, dX .$$

The kinetic energy is the same as in section 4.1.

We obtain

$$\frac{1}{2}\int_{t_1}^{t_2}\int_0^l \{\mu \dot{u}^2 - EI(u'')^2 + P(t)(u')^2\}\, dX\, dt = extr. \quad \text{(L)}$$

(L) might be solved by Ritz's direct method.

Geometric constraints (boundary conditions):

$$u(0,t) = 0 , \quad u'(0,t) = 0$$

Equation (H), again, can be used as basis for Galerkin's method.

Goldstein's form of Hamilton's principle for non-conservative systems (cf. pp. 38/39 in the book cited in section 4.2) is misleading because he defines a work function

$$W = \sum_i \underset{\sim}{F}_i \cdot \underset{\sim}{r}_i \qquad \text{(vector notation)}$$

and

$$\delta W = \sum_i \underset{\sim}{F}_i \cdot \delta \underset{\sim}{r}_i$$

even for $\underset{\sim}{F}_i$ which may depend on $\underset{\sim}{r}_i$, $\dot{\underset{\sim}{r}}_i$, etc.

M. Levinson (Application of the Galerkin and Ritz methods to nonconservative problems of elastic stability, ZAMP 17 (1966), 431-442) who tries to save Ritz's method for nonconservative systems starts from Goldstein's form of Hamilton's principle. I consider his investigations to be somewhat arbitrary and, mathematically, not straight forward.

Cf. H. Leipholz, Über die Befreiung der Ansatzfunktionen des Ritzschen und Galerkinschen Verfahrens von den Randbedingungen, Ing.-Arch. 36 (1967), 251-261.

10. Damping (linear)

10. 1 Linear external damping

To take "some" damping effects into account, the equations of motion, cf. sect. 2, are frequently supplemented by a term $b\dot{u}$; b - damping coefficient,

$$Lu + b\dot{u} + \mu\ddot{u} = 0. \qquad (*)$$

(*) is still linear but, in general, in this form it is not separable (see section 11). By assuming

$$u = U(\underset{\sim}{X})\,T(t) \qquad (A)$$

we obtain from (*)

$$\frac{1}{U}\,LU + b\frac{\dot{T}}{T} + \mu\,\frac{\ddot{T}}{T} = 0 \ . \qquad (B)$$

Separation impossible if b and/or μ depend on the space variables.

10.11 Special separable case : b is proportional to μ

In this case $b = \beta\mu$, β-constant, (C)

and from (B) we obtain

$$L\,U = \lambda\mu\,U \tag{D}$$

$$\ddot{T} + \beta\dot{T} + \lambda T = 0. \tag{E}$$

Additionally we have the boundary conditions (which we assume to be separable). If the eigenvalue problem constituted by (D) and the boundary conditions has the set of (positive) eigenvalues λ_n, $n = 1, 2, \ldots$, then (E) has the solutions

$$T_n = e^{-\beta t/2}\left(A_n \sin\omega_n t + B_n \cos\omega_n t\right),$$

where $\omega_n = \sqrt{\lambda_n - \beta^2/4}\,,$

$$\gamma_n = -\beta/2 \pm i\,\omega_n\,, \qquad \text{cf. section 8.2/8.3.}$$

If β is large, some of the lower γ_n may be negative real).

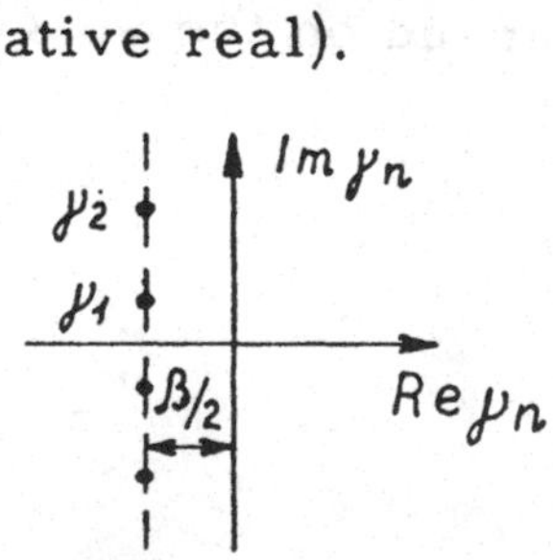

Fig. 10.1

The set of the γ_n is shifted into the negative imaginary half-plane of the complex plane (cf. Fig. 8.3 and Fig. 10.1).

Further separable cases exist for special systems

10. 12 The general case

The assumption

$$u = U(\underset{\sim}{X})\, e^{\gamma t}$$

transforms (*) and the boundary conditions into

$$LU + \gamma bU + \gamma^2 \mu U = 0 \qquad \text{(F)}$$

and boundary conditions for U which equally may include the eigenvalue γ. Such "generalized" eigenvalue problems are sometimes called "nonlinear" because γ occurs non-linear in equation (F) and/or the boundary conditions.

Little general investigations of such problems are known (for instance the notion of adjoint operators, cf. sect. 5, is not helpful in this case; cf. sect. 11). Numerical determination of eigenvalues, γ_n, and eigenfunctions, U_n is possible but tedious (cf. section 13).

In general, both the γ_n and the U_n are complex. The computations can be done with complex numbers, or — after splitting γ, U, the equation (F), the boundary conaitions, and the characteristic equation into real and imaginary parts — with real numbers. The result

$$u(\underset{\sim}{X}, t) = U(\underset{\sim}{X}) e^{\gamma t} \tag{G}$$

is, in general, a complex function of $\underset{\sim}{X}$ and t. <u>But if</u> (*) and the boundary conditions (for u) are real, then

$$\begin{aligned} u_R(\underset{\sim}{X}, t) &= Re\,[U(\underset{\sim}{X}) e^{\gamma t}] \\ &= e^{\delta t}\,[U_R \cos\omega t - U_I \sin\omega t], \end{aligned} \tag{H}$$

$$\begin{aligned} u_I(\underset{\sim}{X}, t) &= Im\,[U(\underset{\sim}{X}) e^{\gamma t}] \\ &= e^{\delta t}\,[U_R \sin\omega t + U_I \cos\omega t], \end{aligned} \tag{I}$$

$$(\gamma = \delta + i\omega),$$

are two linearly independent, real solutions. (The subscript, n, is dropped in these equations.)

The solution (G) can be interpreted visually as shown in Fig. 10.2.

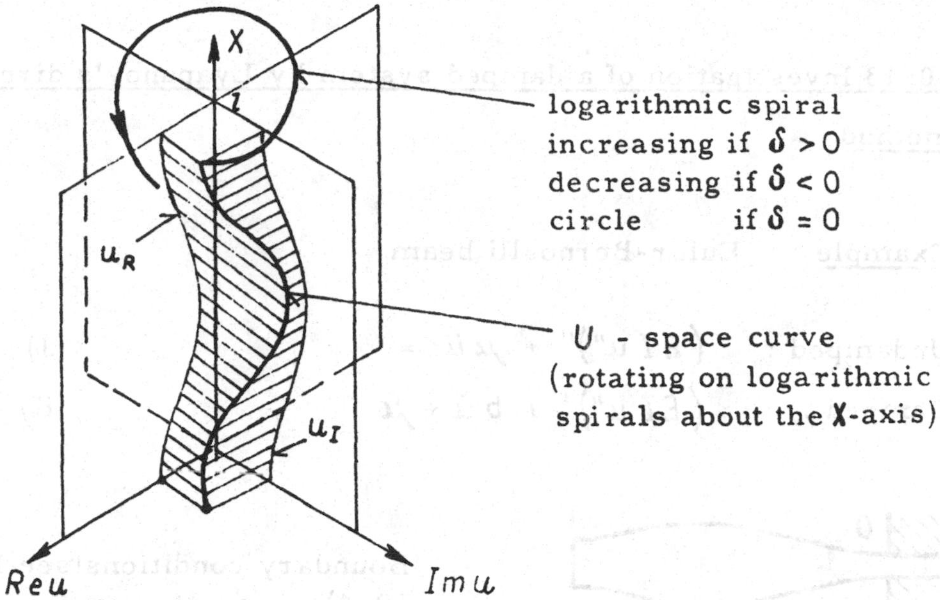

Fig. 10. 2

From equations (H), (I), and from figure 10. 2 it is easily seen that there do not exist fixed nodal points (cf. sect 5. 3) if a (real) separation of the variables is impossible (cf. section 11). The nodal points – the zeroes of u_R , u_I – move along the X - axis.

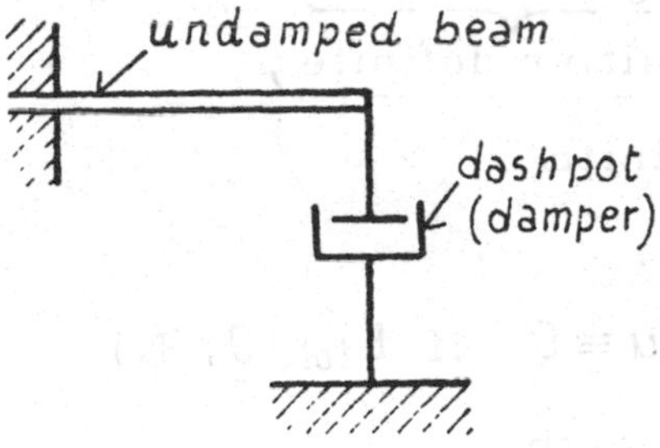

Fig. 10. 3

If the equation of motion of the continuous part of the system is separable but the boundary conditions are not separable (with identically the same parameter of separation) then there do not exist separable (real) solutions of the problem (cf. the system sketched in Fig. 10. 3). The solutions have the form (G).

10. 13 Investigation of a damped system by Lyapunov's direct method

Example : Euler-Bernoulli beam

Undamped : $(EI\,u'')'' + \mu\ddot{u} =$ (J)

Damped : $(EI\,u'')'' + b\dot{u} + \mu$ (K)

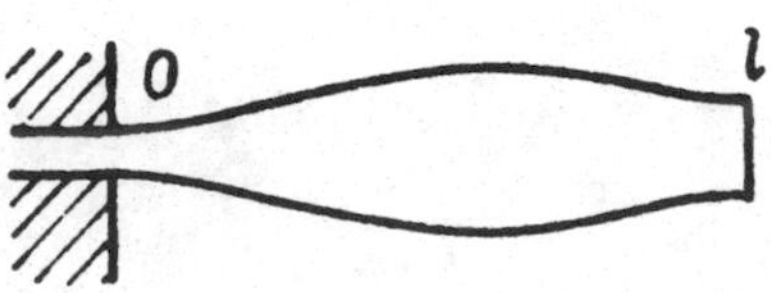

Fig. 10. 4

Boundary conditions (see Fig. 10. 4)

$$\left.\begin{aligned} u(0,t) &= u'(0,t) = 0 \\ u''(l,t) &= (EI\,u'')'\big|_{x=l} = 0 \end{aligned}\right\} \quad (L)$$

Total energy for the undamped system

$$E_{tot} = \underbrace{\underbrace{\frac{1}{2}\int_0^l EI\,(u'')^2 dX}_{\text{positive definite}} + \underbrace{\frac{1}{2}\int_0^l \mu\,u^2\,dX}_{\text{positive definite}}}_{\text{positive definite}} \quad (M)$$

$\hookrightarrow E_{tot} > 0$ if $u \not\equiv 0$ and $u \equiv 0$ if $E_{tot} = 0$, (N)

u sufficiently smooth.

$E_{tot} = V(u)$ is used as Lyapunov's function for the damped and the undamped system.

Differentiation of $V(u)$ with respect to the time

yields

$$\dot{V} = \int_0^l EI\,u''\dot{u}''\,dX + \int_0^l \mu\,\dot{u}\,\ddot{u}\,dX. \qquad (O)$$

By partial integration and substitution of (K) we obtain from (O)

$$\dot{V} = -\int_0^l b\,\dot{u}^2 dX \leq 0,$$

but $\dot{u} \equiv 0$, $u \not\equiv 0$ is no solution of (K), thus, $V \to 0$ and $u(X,t) \to 0$, all disturbances (oscillations) fade away $\longleftrightarrow$ asymptotic stability (cf. sect. 12). (P)

Similar arguments hold if in (K) the term $b\dot{u}$ is replaced by other, e.g., nonlinear, damping terms.

For the stability problem mentioned in the section 8.1 the energy E_{tot} is not positive definite, therefore a conclusion similar to (P) is not possible. Cf. section 2.3 in H. Leipholz, Stabilitätstheorie, Teubner, Stuttgart 1968.

Literature on Lyapunov's method :

W. Hahn, Theorie und Anwendung der direkten Methode von Lyapunov, Springer, Berlin 1959

V.I. Zubov, Methods of A.M. Lyapunov and their application, Noordhoff, Groningen 1964.

10. 14 Forced vibrations

Forced vibrations can be studied in a similar way as in chapter 6. Since the natural frequencies, ω_n , are replaced by the complex values γ_n a resonance catastrophe with deflections of arbitrarily large amplitudes can not occur, because the real frequencies Ω of the exciting forces can never coincide with $\gamma_n = \delta_n + i\omega_n$. Comparatively large amplitudes occur if $|\Omega - \omega_n|$ and $|\delta_n|$ are small.

This subject is treated in many books. Here we mention only E. Skudrzyk's book (written from the acoustical point of view) where much more literature is cited :

E. Skudrzyk, Simple and complex vibratory systems, The Pennsylvania State University Press, 1968.

10. 2 Linear viscoelastic internal damping

The internal damping of the material is frequently taken into account by a viscoelastic stress strain relation :

$$\sigma = E_1 \varepsilon + E_2 \dot{\varepsilon} . \qquad \text{(A)}$$

Constitutive equations of this form are usually visualized by "models" composed of dashpots and springs as shown in Figs. 10. 5, 10. 6.

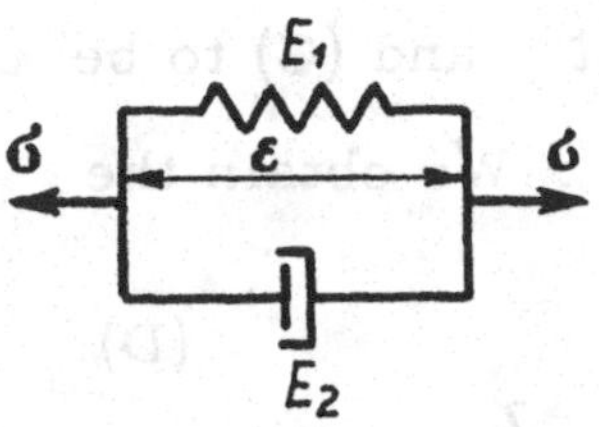

Voigt-Kelvin
($\triangleq$ equ. (A))

Fig. 10. 5

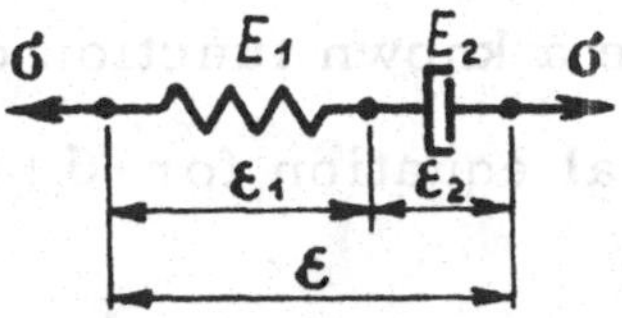

Maxwell
($\triangleq$ equ. (C))

Fig. 10. 6

Various combinations of these elements are possible. They represent constitutive equations of the form

$$\left(a_0 + a_1 \frac{\partial}{\partial t} + a_2 \frac{\partial^2}{\partial t^2} + \ldots\right)\sigma = \left(b_0 + b_1 \frac{\partial}{\partial t} + b_2 \frac{\partial^2}{\partial t^2} + \ldots\right)\varepsilon \quad \text{(B)}$$

If (B) contains time derivatives of σ the momentary stress depends on the strain-history.

Example : The Maxwell model (cf. Fig. 10. 6)

Stress strain relation :

$$\sigma = \varepsilon_1 E_1 = \dot{\varepsilon}_2 E_2 \ , \quad \varepsilon = \varepsilon_1 + \varepsilon_2$$

Elimination of ε_1 , ε_2 leads to

$$\frac{\dot{\sigma}}{E_1} + \frac{\sigma}{E_2} = \dot{\varepsilon} \qquad \text{(C)}$$

We consider $\varepsilon(t)$ to be a known function of t and (C) to be an ordinary differential equation for $\sigma(t)$. We obtain the solution

(D)

$$\sigma(t) = E_1 \int_{-\infty}^{t} e^{-E_1(t-\tau)/E_2} \dot{\varepsilon}(\tau)\,d\tau = E_1\varepsilon(t) - E_2 \int_{-\infty}^{\tau} e^{-E_1(t-\tau)/E_2} \varepsilon(\tau)\,d\tau .$$

Lit. : Short summary : Chapter 53 (by E. H. Lee) in W. Flügge's Handbook cited in section 2.242

Extension to three-dimensional states of stress : S. Flügge, Handbuch der Physik, Bd. 6, Elastizität und Plastizität, Springer, Berlin 1958

10.21 Distribution of eigenvalues

Example : Euler-Bernoulli beam with stress strain relation (A).

μ, I, E_1, E_2 - constant

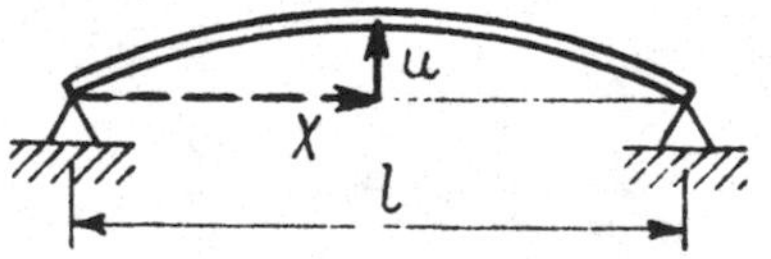

Fig. 10.7

Equation of motion :

$$E_1 I u^{IV} + E_2 I \dot{u}^{IV} + \mu \ddot{u} = 0 \qquad \text{(E)}$$

Boundary conditions :

$$u(0,t) = u(l,t) = 0$$

$$M(0,t) = E_1 I u''(0,t) + E_2 I \dot{u}''(0,t) = 0$$

$$M(l,t) = E_1 I u''(l,t) + E_2 I \dot{u}''(0,t) = 0$$

Both the equation of motion and the boundary conditions are separable.

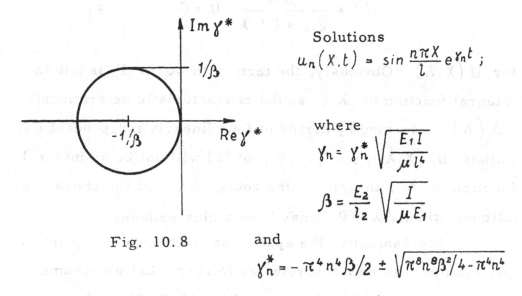

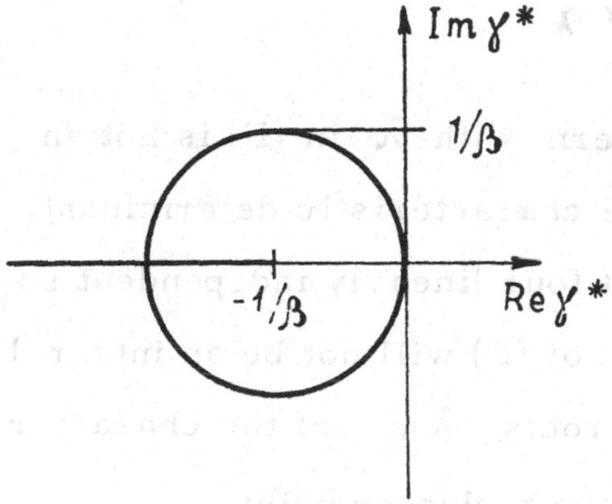

Fig. 10.8

Solutions

$$u_n(X,t) = \sin\frac{n\pi X}{l}\, e^{\gamma_n t};$$

where

$$\gamma_n = \gamma_n^* \sqrt{\frac{E_1 I}{\mu l^4}}$$

$$\beta = \frac{E_2}{l_2}\sqrt{\frac{I}{\mu E_1}}$$

and

$$\gamma_n^* = -\pi^4 n^4 \beta/2 \pm \sqrt{\pi^8 n^8 \beta^2/4 - \pi^4 n^4}$$

The couples of complex conjugate values γ_n^* lie on the circle shown in Fig. 10.8, the real values γ_n^* lie in the interval $(-\infty, -1/\beta)$ of the real axis, $-1/\beta$ is a cluster point.

Remark concerning the cluster point (cf. section 5.2):

Assuming a solution of the form

$$u(X,t) = U(X)\, e^{\lambda t}$$

for (E) we obtain the differential equation

$$U^{IV} + \frac{\mu \lambda^2}{E_1 I + E_2 I \lambda}\, U = 0 \qquad \text{(F)}$$

for $U(X,\lambda)$. Obviously, the term with λ in (F) is not an integral function of λ, so the characteristic determinant, $\Delta(\lambda)$, formed by means of four linearly independent solutions $U_i(X,\lambda)$, $i = 1, \cdots, 4$, of (F) will not be an integral function of λ, therefore, the roots λ_n of the characteristic equation, $\Delta(\lambda) = 0$, may have a cluster point.

Mechanically, the appearance of the cluster point in our example can be interpreted as follows : Let us assume the beam has initially a kink, see Fig. 10.9. Since the material is viscoelastic the kink will disappear <u>slowly</u>. Thus, to represent this behavior by a Fourier expansion with respect to the eigenfunctions, cf. sect. 5.5,

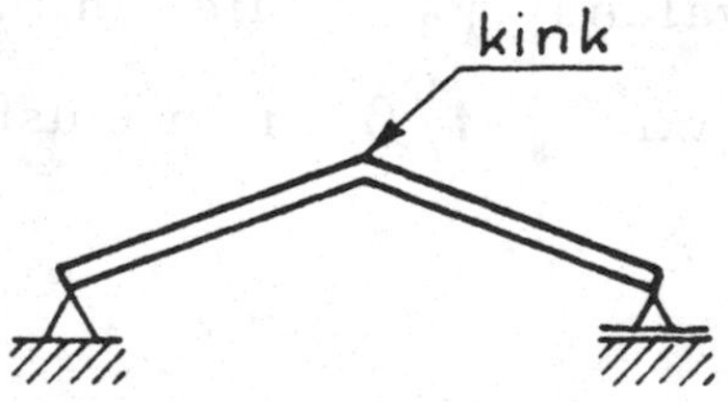

Fig. 10.9

we need high order eigenfunctions - to be able to represent the kink -which belong to eigenvalues with comparatively small negative real parts. Such functions are provided by the eigenvalues in the vicinity of the cluster point.

10. 22 Self-excited vibrations of a rotating shaft caused by internal damping

Internal damping can be the cause of self-excited vibrations of a rotating shaft.

Lit. : W. Kellenberger, Die Stabilität rotierender Wellen infolge innerer und äusserer Dämpfung, Ing. -Arch. 32 (1963) p. 323

G. F. Schirmer, Zur Stabilität der Schwingungen von Turbinenwellen, Dissertation, Techn. Hochschule Darmstadt, 1969.

Survey article : R. E. D. Bishop, G. Parkinson, Vibration and balancing of flexible shafts, Appl. Mech. Rev. 21, No. 5 (1968).

10. 3 "Complex modulus" damping

The hysterctic loop of a Voigt-Kelvin material(cf. Fig. 10. 5 and equation (A) in sect. 10. 2),extended sinusoidally,is an ellipse, see Fig. 10. 10. Metals show such

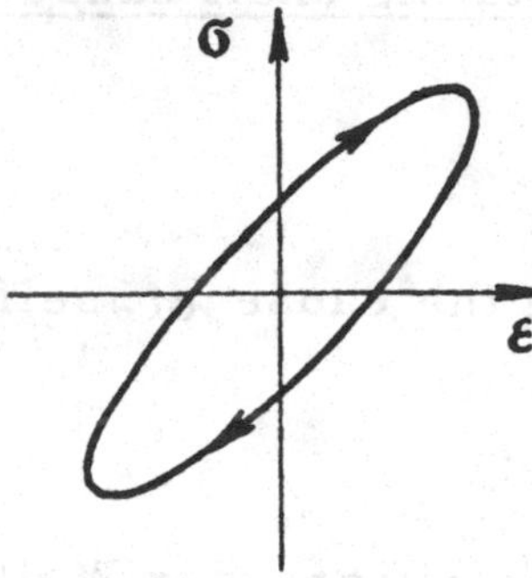

Fig. 10. 10

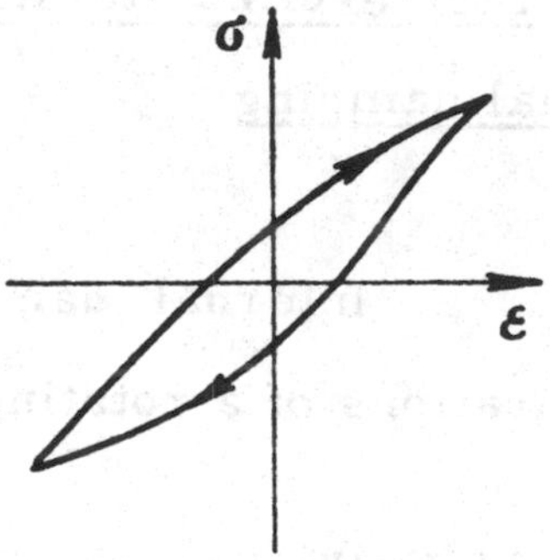

Fig. 10. 11

a behavior only at low stresses. For intermediate stresses metals show a rate-independent damping and a hysteretic loop of the form indicated in Fig. 10. 11 (nonlinear!).

For practical problems the rate-independence of the damping is often approximated by the linear relation

$$\sigma = E^{*}\varepsilon , \tag{G}$$

where E^{*} is the complex modulus of elasticity

$$E^{*} = E' + i E'' , \quad i = \sqrt{-1}$$

E' - storage modulus (elastic modulus, real modulus)

E" - loss modulus (dissipation modulus)
$E''/E' = \eta$ - loss coefficient (loss factor, damping factor)

In the relation (G) σ and ε are assumed to be complex and to be represented by complex Fourier series or Fourier integrals. (So the actual values are $Re\,\sigma$ and $Re\,\varepsilon$, or $Im\,\sigma$ and $Im\,\varepsilon$). Correspondingly, equations of motion in which (G) is involved must be interpreted and solved in an adequate way.

Frequently, for special systems, E^* is obtained by sinusoidal excitation of the system and given in one of the forms $E^*(\omega)$ or $E^*(\omega, A)$ where ω is the frequency and A is a characteristic amplitude. The application of such an E^* to non-sinusoidal motions, they might even be periodic, is not possible from the logical standpoint.

Lit. : R. Plunkett, Damping definitions and classifications, Lecture at the 1968 meeting of the Acoustical Society of America in Cleveland, Ohio.

Survey articles in "Applied mechanics surveys", ed. by Abramson, Liebovitz, Crowley, Juhasz, Spartan Books, Washington, D. C., 1966:

R. Plunkett, Vibration Damping, pp. 691-702,

B. J. Lazan, Damping properties of materials, members and composites, pp. 703-716.

11. Some remarks on the separability of damped systems and non-self-adjoint eigenvalue problems

11. 1 Transformation of a "damped equation of motion" into a separable form

We introduce

$$\left.\begin{aligned} u_1(\underset{\sim}{X},t) &= u(\underset{\sim}{X},t) \\ u_2(\underset{\sim}{X},t) &= \dot{u}(\underset{\sim}{X},t) \end{aligned}\right\} \underset{\sim}{u} = \begin{pmatrix} u_1 \\ u_2 \end{pmatrix} \tag{A}$$

into the equation (*) of section 10. 1 and obtain

$$\underset{\sim}{L}\,\underset{\sim}{u} = \underset{\sim}{\mu}\,\dot{\underset{\sim}{u}}, \tag{B}$$

where

$$\underset{\sim}{L} = \begin{pmatrix} 0 & 1 \\ -L & -b \end{pmatrix}, \quad \underset{\sim}{\mu} = \begin{pmatrix} 1 & 0 \\ 0 & \mu \end{pmatrix}.$$

Equation (B) can be separated formally :

$$\underset{\sim}{u}(\underset{\sim}{X},t) = \underset{\sim}{U}(\underset{\sim}{X})\,T(t), \tag{C}$$

$$\underset{\sim}{L}\,\underset{\sim}{U} = \gamma\,\underset{\sim}{\mu}\,\underset{\sim}{U}, \tag{D}$$

$$\dot{T} = \gamma T. \tag{E}$$

In general the eigenvalue problem constituted by (D) and a proper set of boundary values will not be self-adjoint (the eigenvalues γ_n need not be real, cf. sect. 10. 11).

11. 2 Non-self-adjoint eigenvalue problems

Example : Damped longitudinal vibrations of a rod, cf. section 2. 1 and Fig. 11. 1.

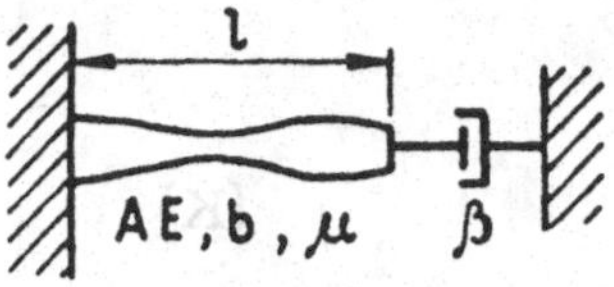

Fig. 11. 1

Equation of motion :

$$-(AEu')' + b\dot{u} + \mu\ddot{u} = 0 . \tag{F}$$

Boundary conditions :

$$\left.\begin{aligned} x = 0 &: u = 0 \\ x = l &: AEu' + \beta\dot{u} = 0 . \end{aligned}\right\} \tag{G}$$

The transformation (A) does not yield a form suitable for the further investigation. Therefore, we replace (F) by the first order system

$$\underset{\sim}{L}\,\underset{\sim}{u} = \underset{\sim}{\mu}\,\dot{\underset{\sim}{u}} \tag{H}$$

where

$$u_1 \equiv u \ , \quad u = \begin{pmatrix} u_1 \\ u_2 \end{pmatrix} ,$$

$$\underset{\sim}{L} = \begin{pmatrix} 0 & 1 \\ AE & 0 \end{pmatrix} \frac{\partial}{\partial x} - \begin{pmatrix} b & 0 \\ 0 & 0 \end{pmatrix} , \quad \underset{\sim}{\mu} = \begin{pmatrix} \mu & 0 \\ 0 & 1 \end{pmatrix} .$$

The boundary conditions (G) are equivalent to

$$x = 0 : \ u_1 = 0 \ , \quad x = l : \ \dot{u}_2 + \beta\dot{u}_1 = 0 . \tag{I}$$

(Some general considerations about the equivalence of a system of differential eqauations and a single differential equation see chapt. I, § 2 in vol. 2 of Courant, Hilbert's book cited in sect. 5. 2).

The separation (C) leads to the eigenvalue problem

$$\underset{\sim}{L}\,\underset{\sim}{U} = \gamma \underset{\sim}{\mu}\,\underset{\sim}{U}\ , \quad \underset{\sim}{\ell}\,\underset{\sim}{U} = 0\ . \tag{J}$$

$\underset{\sim}{\ell}\,\underset{\sim}{U} = 0$ denotes the boundary conditions

$$U_1(0) = 0\ , \quad U_2(l) + \beta U_1(l) = 0\ . \tag{K}$$

By the procedure outlined in section 5.42 we obtain the adjoint eigenvalue problem

$$\underset{\sim}{L}^{+}\,\underset{\sim}{V} = \nu \underset{\sim}{\mu}\,\underset{\sim}{V}\ , \quad \underset{\sim}{\ell}^{+}\underset{\sim}{V} = 0\ , \tag{J$^+$}$$

where $\underset{\sim}{L}^{+} = -\begin{pmatrix} 0 & AE \\ 1 & 0 \end{pmatrix}\dfrac{\partial}{\partial x} - \begin{pmatrix} b & (AE)' \\ 0 & 0 \end{pmatrix} \neq \underset{\sim}{L}\ ,$

$$\underset{\sim}{\ell}^{+}\underset{\sim}{V} = 0:\ V_1(0) = 0\ ,\ AE\,V_2(l) = \beta V_1(l)\ , \quad \underset{\sim}{\ell} \neq \underset{\sim}{\ell}^{+}$$

The eigenvalue problem (J) is not self-adjoint.

Neumark (cited in sect. 5.2) proves : If γ is an eigenvalue of (J) then $\nu = \bar{\gamma}$ (complex conjugate) is an eigenvalue of (J^+). But, in general, if $\underset{\sim}{U}$ is an eigenfunction of (J), $\underset{\sim}{V} = \bar{\underset{\sim}{U}}$ need not be an eigenfunction of (J^+).

11.21 Orthogonality of eigenfunctions

Let $\underset{\sim}{U}$, λ, and $\underset{\sim}{V}$, ν be eigenfunctions and the correlated eigenvalues of (J) and (J^+), respectively. Similarly as in section 5.44 we obtain (for the one dimensional case)

$$\underbrace{(\underset{\sim}{L}\,\underset{\sim}{U}\,,\underset{\sim}{V}) - (\underset{\sim}{U}\,,\underset{\sim}{L}^{+}\underset{\sim}{V})}_{=0} = \underbrace{(\gamma - \bar{\nu})}_{\substack{\neq 0 \\ \text{if } \gamma \neq \bar{\nu}}} \underbrace{\int_0^l \underset{\sim}{U}^T \underset{\approx}{\mu}^T \bar{\underset{\sim}{V}}^T \, dX}_{\hookrightarrow = 0} .$$

$\underset{\sim}{U}$ and $\underset{\sim}{V}$ are orthogonal if $\gamma \neq \bar{\nu}$:

$$\int_0^l \underset{\sim}{U}^T \underset{\approx}{\mu}^T \bar{\underset{\sim}{V}} \, dX = 0 \qquad \text{(L)}$$

If the eigenfunctions $\underset{\sim}{U}$ and $\underset{\sim}{V}$ are known, (L) can be applied in the same way as the corresponding relation in section 5. 44 to solve the initial value problem, cf. section 5. 5.

The notions mentioned in this section are closely related to corresponding notions in the theory of integral relations, cf. Courant, Hilbert, vol. I;

W. A. Smirnov, Lehrgang der höheren Mathematik, Bd. IV, Deutsch. Verl. d. Wissensch., Berlin.

12. Lyapunov's definition of stability

Let

$$G[u] = 0 \qquad \text{(A)}$$

be an equation of motion which governs the motion of a given system completely (differential equation with proper boundary conditions, Hamilton's principle in its general form, etc.; cf. sect. 9). In general u will be an N-dimensional vector, cf. $\underset{\sim}{x}(\underset{\sim}{X}, t)$ in section 1.2.

Let

$$u = u^*(\underset{\sim}{X}, t)$$

be a solution of (A). We define a new state variable $v(\underset{\sim}{X}, t)$ by

$$v = u - u^* \qquad \text{(B)}$$

Introducing v into (A) we obtain

$$G^*[v] = 0, \qquad (\mathrm{A}^*)$$

an equation for $v(\underset{\sim}{X}, t)$

In ordinary differential equations the -nonlinear-equation (A^*) is sometimes called the variational equation - belonging to the solution u^* -(cf. L. Cesari, Asymptotic behavior and stability problems in ordinary differential equations,

Springer, Berlin 1963]. But in general the linearization of (A^*) with respect to v ,

$$G_L^*[v] = 0, \qquad (A_L^*)$$

is called variational equation. (Cesari calls (A_L^*) linearized variational equation).

u^* is called the undisturbed motion of the given system and v is the disturbance. (A^*) and (A_L^*) have always the trivial solution $v^* \equiv 0$ which corresponds to the solution u^* of (A).

Let

$$\| u \| \geqslant 0 \qquad (C)$$

be a norm for all u which are elements of a proper function space (cf. $\|U\|$ in sect. 5.41), and

$$\| u \| = 0 \quad \text{if and only if} \quad u \equiv 0 \qquad (D)$$

(Some more relations which $\|u\|$ has to satisfy and some generalizations see Zubov's book cited in sect. 10.13).

<u>Example</u>: Scalar $u(X,t)$, $0 \leq X \leq l$

$$\| u \| = \left[\int_0^l u^2 \, dX \right]^{1/2} - L^2\text{-norm}.$$

12. 1 Lyapunov's (weak) stability

The trivial solution $v^* \equiv 0$ of (A^*) - and, correspondingly, the solution u^* of (A) - is callled (weakly) stable if for any $\varepsilon > 0$ there exists a $\delta > 0$ such that

$$\| v(\underset{\sim}{X}, t) \| < \varepsilon \quad for \quad t \geq t_0$$

if

$$\| v_0 \| < \delta \; , \quad \delta = \delta(\varepsilon) \; ,$$

where $v_0 = v(\underset{\sim}{X}, t_0)$ is the initial state (disturbance)at $t = t_0$.

12. 2 Lyapunov's asymptotic stability

The trivial solution is called asymptotically stable if it is stable and if

$$\| v(\underset{\sim}{X}, t) \| \longrightarrow 0 \quad for \quad t \longrightarrow \infty .$$

12. 3 Lyapunov's unstability

The trivial solution is called unstable if it is not (weakly) stable, cf. 12. 1.

12. 4 Remarks

The stability definitions given here concern disturbances of the initial conditions (12. 1:"The disturbed solution $v(\underset{\sim}{X},t)$ stays in an ε-neighborhood of the trivial solution, $v^* \equiv 0$, if the initial disturbance , v_0 , is small enough!") Extensions to disturbances of parameters see Hahn's book cited in section 10. 13.

There exist many more stability definitions, see, for instance, J. LaSalle, S. Lefschetz, Stability by Lyapunov's second method with applications, Academic Press, New York 1961

W Bogusz, Lyapunov's stability in engineering, pp. 61-67 in Zagadnienia Drgan Nieliniowych (Nonlinear Vibration Problems), Vol. 9, Polish Academy of Sciences, Warszawa 1968.

12. 5 Stability investigation by means of the (linear) variational equation

In many cases the stability of a given system is investigated on the basis of the linear equation (A_L^*) because its solutions are much easier to discuss than the solutions of the nonlinear equation (A^*).

For ordinary differential equations and equations of motion, $G^*[v] = 0$, which can be split into

$$G^*[v] = G_L^*[v] + G_2^*[v] = 0, \tag{E}$$

where $G_2^*[v]$ is of second or higher order with respect to v, Lyapunov proved the following theorems :

<u>12.51</u> If the trivial solution of the (linearized) equation (A_L^*) is <u>asymptotically</u> stable then the trivial solution of the nonlinear equation (A^*) will be asymptotically stable.(For sufficiently small disturbances).

<u>12.52</u> If the trivial solution of the (linearized) equation (A_L^*) <u>is unstable</u> then the trivial solution of the nonlinear equation (A^*) will be unstable.

<u>12.53</u> If the trivial solution of the (linearized) equation (A_L^*) is only <u>weakly</u> stable then the stability of the trivial solution of the nonlinear equation (A^*) depends on the higher order terms, it may be either stable or unstable. (This is the so-called "critical case").

To my knowledge there do not exist the corresponding theorems, e.g., for partial differential equations (continuous systems), but I feel sure that similar relations hold in that field too.

Let us reflect upon the common stability problems of civil engineering from this point of view: Most stability problems civil engineers investigate concern linear conservative systems. In the case of stability the eigenvalues, γ, are distributed along the imaginary axis, cf. sect. 8. If some damping, which is always found in an actual system, is taken into account, the eigenvalues are shifted into the left half of the complex plane, cf. sects. 10.11, 10.21, the trivial solution becomes asymptotically stable. Hence, small nonlinear terms can not change the type of stability.

13. Kinetic stability of autonomous systems

A system is called autonomous if its equation of motion does not depend explicitly on the time, otherwise it is called non-autonomous.

A mechanical system is in the state of static equilibrium under a given constant load. At the time $t = 0$ the system is disturbed (e.g., displaced slightly from its equilibrium state). What happens to the system for $t > 0$? The figures 13.1-13.6 show six different types of response.

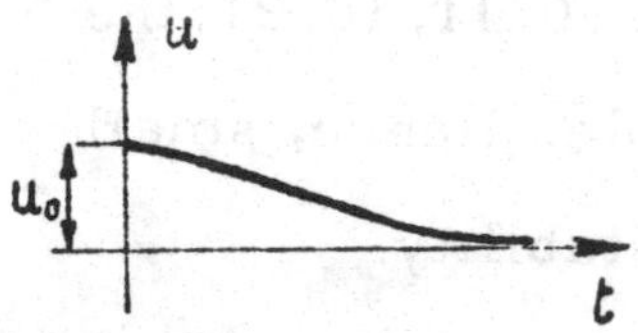

Fig. 13.1

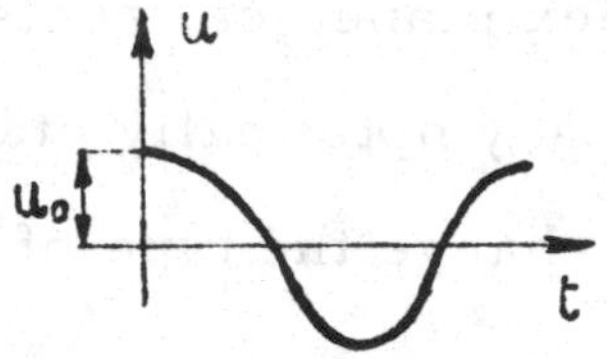

Fig. 13.3

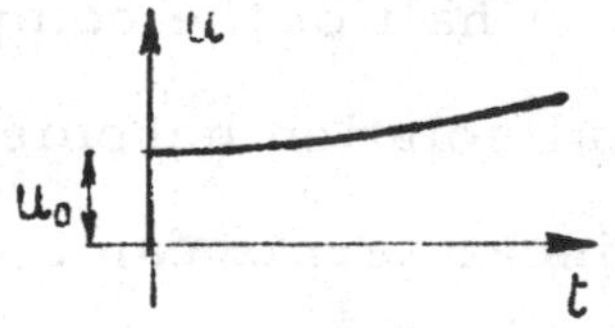

Fig. 13.5

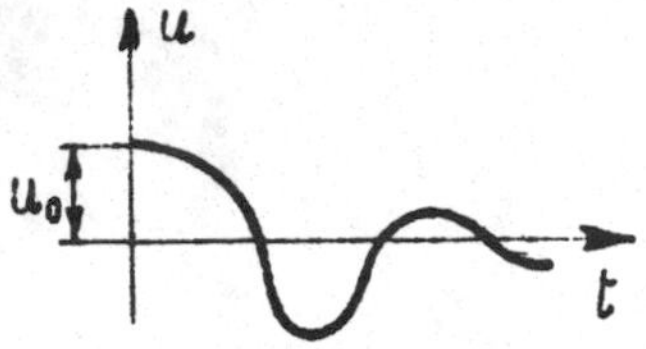

Fig. 13.2

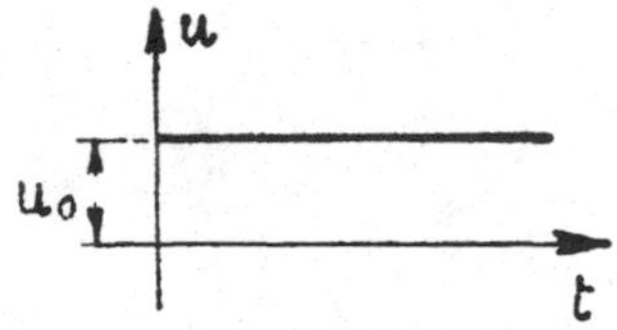

Fig. 13.4

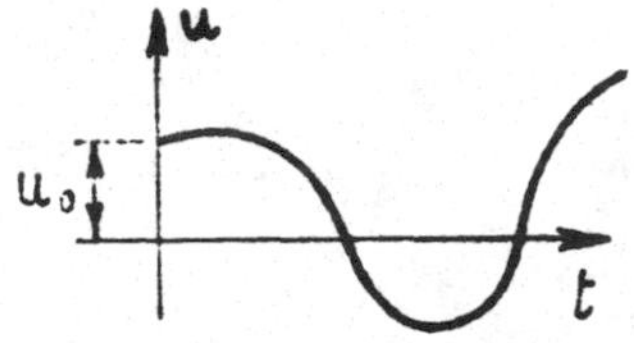

Fig. 13.6

By u we denote the deviation from the equilibrium position, u_0 is a small initial distrubance.

In Fig. 13.1 and Fig. 13.2 the disturbances disappear "exponentially" and "oscillatory", respectively, the system returns to its original equilibrium position. In Fig. 13. 3 the disturbance causes an oscillation with amplitudes which have the magnitude of u_0 . In Fig. 13. 4 the "disturbed" position is a new equilibrium position. In Fig. 13. 5 and Fig. 13. 6 the initial disturbance causes deviations which increase exponentially and oscillatory, respectively.

Corresponding to Lyapunov's definition we have :

Fig.	13. 1	13. 2	13. 3..	13. 4	13. 5	13. 6
stability	asymptotic		weak		unstable	

Fig. 13. 7

We are interested in the unstable cases shown in Fig. 13. 5 and Fig. 13. 6.

The Fig. 13. 5 represents, for instance, the buckling of a column, cf. section 8. The instability starts when the first eigenvalue, γ_1 , moves through the origin, $\gamma = 0$, onto the positive real axis, cf. Fig. 8. 4. The critical load, P_{crit} in section 8, can be found by static methods.

The unstable motion shown in Fig. 3. 6 starts when one of the eigenvalues, γ_{n*} , say, moves into the positive real half-plane, $Re\ \gamma_{n*} > 0$, because the load para-

meter, say P again, is increased over a critical value, P_{crit}
So we have the ("kinetic") stability criterion :

Asymptotic stability if $Re\,\gamma_n < 0$ for all γ_n.

Instability if $Re\,\gamma_n > 0$ for at least one γ_n

The case of weak stability corresponds to the border case $P = P_{crit}$ and $Re\,\gamma_n^* = 0$; $Re\,\gamma_n < 0$ for $\gamma_n \neq \gamma_n^*$.

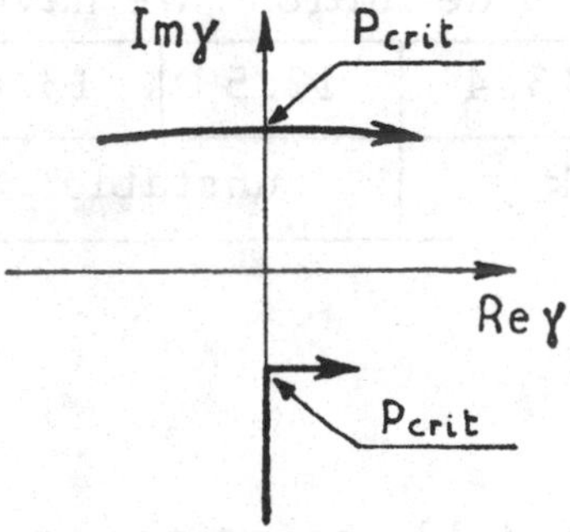

Fif. 13.8

In Fig. 13.8 the arrows indicate two different ways in which an eigenvalue, γ_n^* , can enter the positive real half-plane when P is increased over the critical value, P_{crit}

We see : To find P_{crit} for such systems the eigenvalues γ_n themselves must be investigated. Thus we have to study the motion, i.e., the "kinetic" (or"dynamic") behavior of the system. That is the reason why such problems are called "kinetic" (or "dynamic") stability problems.

In aerodynamics where many of these problems originate the unstable motions from Fig. 13.5 and 13.6 are called "divergence" and "flutter", respectively. We shall use these notations. (Cf. G. Herrmann, R. W. Bungay, On the the stability of elastic systems subjected to nonconservative forces, J. Appl. Mech. 31 (1964) 435-440).

13. 1 Follower forces

Example : Euler-Bernoulli beam loaded as shown in Fig. 13.9; P - follower force of constant magnitude, direction : $\psi = (1 - m)\,\alpha$ (see Fig. 13.9), m - parameter $m = 1$ - vertical P, $m = 0$ - tangential P

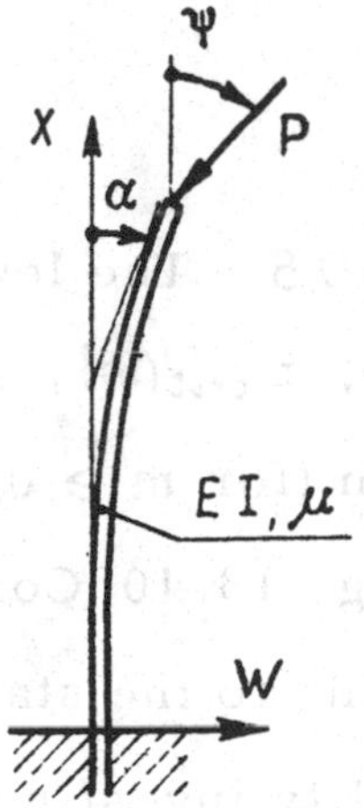

Fig. 13.9

Equation of motion (in a non-dimensional form) :

$$u^{IV} + 2pu'' + \ddot{u} = 0 \qquad \text{(A)}$$

$$u = w/l,\ \xi = X/l,\ \tau = t\sqrt{EI/\mu l^4},$$
$$2p = Pl^2/EI.$$

Boundary conditions

$$\left.\begin{aligned} u(0,\tau) &= u'(0,\tau) = 0 \\ u''(1,\tau) = 0\ ,\ & u'''(1,\tau) + 2pmu'(1,\tau) = 0 \end{aligned}\right\} \text{(B)}$$

The equation (A) is self-adjoint but the boundary conditions, (B), are not if $m \neq 1$.

13.11 Static investigation

The eigenvalue problem reduces to

$$u^{IV} + 2pu'' = 0,$$

and the boundary conditions (B); p is the eigenvalue parameter. We obtain the characteristic equation.

$$m\left(\cos\sqrt{2p} - 1\right) + 1 = 0 \tag{C}$$

(C) has real solutions, p_n, only for $m \geq 0.5$. The lowest value, $p_{crit}(m)$, is shown (for $m \geq 0.5$) in Fig. 13.10. Corresponding to the static stability investigation there would be no stability loss for arbitrarily large loads in the

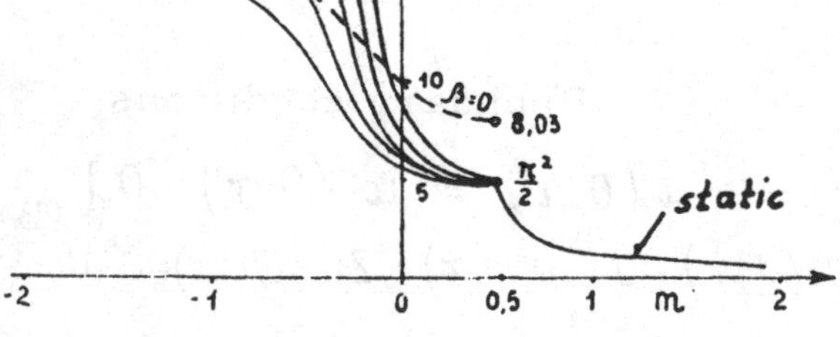

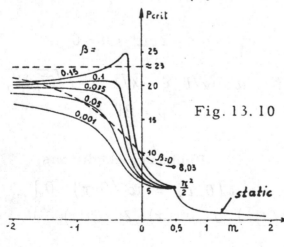

Fig. 13.10

range $m < 0{,}5$.

13.12 Kinetic investigations

Assuming solutions of the form

$$u(\xi,\tau) = U(\xi)\, e^{\gamma\tau}$$

we obtain from (A), (B) the eigenvalue problem

$$U^{IV} + 2pU'' + \lambda U = 0 \;, \quad \lambda = \gamma^2,$$

$$U(0) = U(1) = U''(1) = U'''(1) + 2pm\,U'(1) = 0$$

The characteristic equation can be established in an explicit form (cf. H. König, Die Knickkraft beim einseitig eingespannten Stab unter nichtrichtungstreuer Kraftwirkung, Der Stahlbau 29 (1960) 150-154).

We have the stability condition

$$\operatorname{Re}\gamma_n < 0 \quad \text{for all} \quad \gamma_n$$

The critical load, p_{crit}, is obtained from

$$\operatorname{Re}\gamma_1 = 0 \,,$$

assuming that the lowest normal mode becomes instable first. Unfortunately, the calculations have to be done numerically (cf. H. König cited above). The curve $p_{crit}(m)$ is represented in Fig. 13.10 by a dashed line (marked $\beta = 0$). Surprisingly there is a jump in p_{crit} at $m = 0.5$. (Cf. similar jumps in Hermann & Bungay's paper cited above).

13.13 Influence of damping

König showed that external viscid damping, cf. section 10.1, changes the kinetic stability behavior just slightly and the static stability behavior not at all.

I myself investigated the behavior of a viscoelastic beam loaded as shown in Fig. 13.9 (E. Brommundt, Über den Einfluss von Werkstoffdämpfung auf die Stabilitätsgrenze von Stäben mit Folgelasten, ZAMM 46 (1966) T 117-119):

Equation of motion

$$u^{IV} + \beta \dot{u}^{IV} + 2pu'' + \ddot{u} = 0, \qquad \text{(D)}$$

β - damping coefficient.

Boundary conditions

$$u(0,\tau) = u'(0,\tau) = u''(1,\tau) + \beta \dot{u}''(1,\tau) = u'''(1,\tau) + \beta \dot{u}'''(1,\tau) + 2pmu' = 0. \qquad \text{(E)}$$

The characteristic equation can be established explicitly again, p_{crit} has to be calculated numerically. The results are shown in Fig. 13.10 for some values $\beta > 0$. The curve $\beta \to 0$ is located slightly below the curve $\beta = 0{,}001$.

We see :

1. Damping can decrease p_{crit} in the range of the kinetic stability criterion (cf. H. Ziegler, Die Stabilitätskriterien der Elastomechanik, Ing. -Arch. 20 (1952) 49-56).
2. p_{crit} for $\beta \to 0$ and p_{crit} for $\beta = 0$ need not coincide.
3. In the kinetic range p_{crit} depends on the type of the damping.
4. The jump of p_{crit} at $m = 0.5$ disappears.

In the range of the kinetic stability criterion the mass distribution, too, can influence the stability behavior of the system (cf. Leipholz's book cited in section 10.13). If the load acting on the end of the beam is distributed in a more general way,different types of flutter are possible : see S. Nemat-Nasser, G. Herrmann, Torsional instability of cantilevered bars subjected to nonconservative loading, J. Appl. Mech. (1966) 102-104.

13.14 Movements of the eigenvalues for $\beta = 0$.

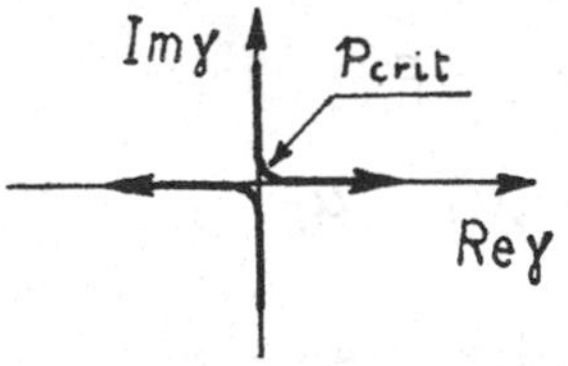

Fig. 13.11

Fig. 13.11 shows how the γ_n move, when p is increased, in the static range, cf. Fig. 8.3

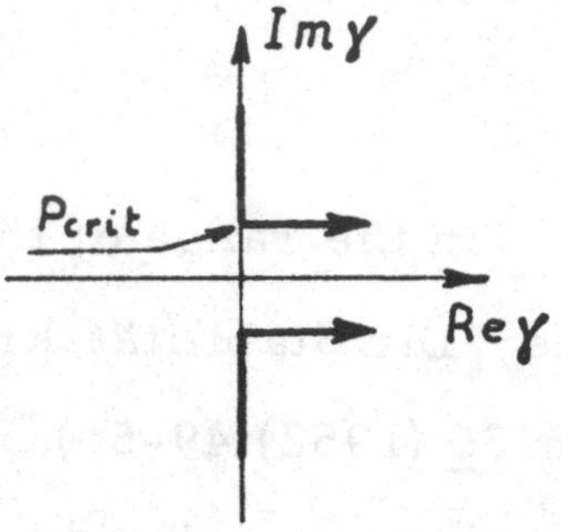

Fig. 13.12

Fig. 13.12 shows how the γ_n move in the kinetic range.

(How do the γ_n move for $\beta \neq 0$?)

13.2 Pipes conveying fluid

<u>Example</u> : Uniform pipe, cf. Fig. 13.13, length l

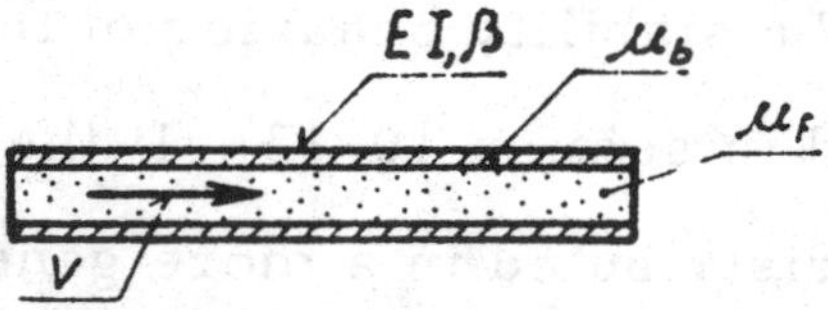

Fig. 13.13

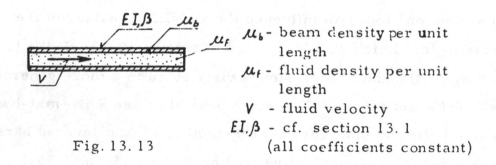

μ_b - beam density per unit length
μ_f - fluid density per unit length
V - fluid velocity
EI, β - cf. section 13.1 (all coefficients constant)

Equation of motion

$$u^{IV} + \beta \dot{u}^{IV} + 2pu'' + q\dot{u}' + \ddot{u} = 0 \tag{A}$$

u, β , etc. cf. section 13.1, $2p = +\mu_f v^2 l^2 / EI$,

$$q = 2 v \mu_f l / \sqrt{\mu EI} , \quad \mu = \mu_b + \mu_f .$$

Additionally, we have four boundary conditions.

In general,this system is not conservative since energy can be exchanged between the transversal oscillations of the pipe and the longitudinal motion of the fluid Therefore,the kinetic criterion must be applied.

Some results, for special boundary conditions, are given,e. g., in the paper S. Nemat-Nasser, S. N. Prasad, G. Herrmann, Destabilizing effect of velocity-dependent forces in nonconservatice continuous systems, Northwestern University, Evanston, Ill. , Techn. Report No. 65-4, and in my paper mentioned above.

13. 21 Cases for which a static stability investigation leads to the proper results

In section 13. 3 we shall refer to Leipholz's more systematic investigations of the problem mentioned in the title. Here we shall derive just a special result obtainable by Lyapunov's method.

Let us define for (A) a Lyapunov function :

$$V(u) = E_{Kin} + E_{pot},$$

where

$$E_{kin} = \frac{1}{2}\int_0^1 \dot{u}^2 d\xi ,$$

$$E_{pot} = \frac{1}{2}\int_0^1 u''^2 d\xi - p\int_0^1 u'^2 d\xi .$$

Differentiating V with respect to τ and eliminating $\ddot{u}$ by means of (A) we get after some calculations

$$\dot{V} = -\beta\int_0^1 (\dot{u}'')^2 d\xi + B ,$$

where

$$B = \left\{ -\dot{u}\,u''' + \dot{u}'u'' - \beta\,\dot{u}\,\dot{u}''' + \beta\,\dot{u}'\dot{u}'' - 2p\dot{u}u' - q\dot{u}^2/2 \right\}\Big|_0^1$$

$\dot{V}$ is not positive if the boundary terms, B, vanish. Then we have stability <u>if</u> V itself is positive definite (cf. Zubov, p. 38). E_{kin} is always positive definite, E_{pot} is positive definite if $p < p_{crit,\,stat}$, where $p_{crit,\,stat}$ is the static critical load. Thus, for $B = 0$ the static stability investigation leads to the proper results. (The condition $B = 0$ is only sufficient but need not be necessary for the admissibility of the static criterion). Follower loads may be included in (A).

13.3 Galerkin's method

Example : Cantilever pipe, conveying fluid (no damping).

Equation of motion(cf. sect. 13.2):

$$u^{IV} + 2pu'' + q\dot{u}' + \ddot{u} = 0 \quad \text{(A)}$$

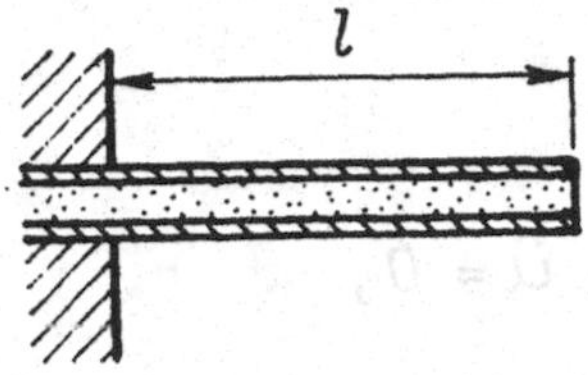

Fig. 13.14

Boundary conditions :

$$u(0,\tau) = u'(0,\tau) = u''(1,\tau) = u'''(1,\tau) = 0. \quad \text{(B)}$$

Introduction of (B) into the expression for the boundary terms, B, in the preceding section does not show that we are allowed to apply the static stability criterion.

By

$$u = U(\xi)\, e^{\gamma\tau}$$

we obtain from (A), (B) the eigenvalue problem

$$U^{IV} + 2pU'' + \gamma q U' + \gamma^2 U = 0\,, \quad \text{(C)}$$

$$U(0) = U'(0) = U''(1) = U'''(1) = 0. \quad \text{(D)}$$

We apply Galerkin's method to solve (C), (D).

As approximate functions we use the first N normal modes of the cantilever beam shown in Fig. 13.15. Its equation of motion is

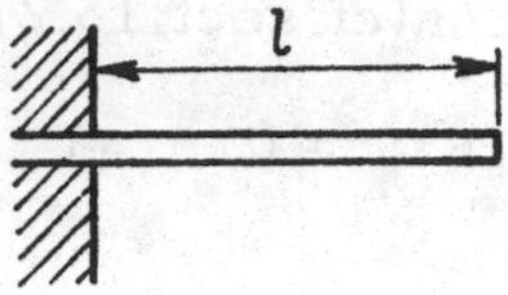

Fig. 13.15

$$u^{IV} + \ddot{u} = 0,$$

the boundary conditions are the same as in Fig. 13.14, i.e., the conditions (B). We denote the normal modes of the cantilever beam by V_n

Introducing the approximation

$$U = \sum_{n=1}^{N} a_n V_n \tag{E}$$

into (C) and applying the Galerkin projection method (cf. sect. 7) we obtain a set of N linear homogeneous equations for the coefficients a_n . In matrix notation we have

$$\left(\underset{\sim}{A} + \gamma \underset{\sim}{B} + \gamma^2 \underset{\sim}{I} \right) \underset{\sim}{a} = 0 \tag{F}$$

$\underset{\sim}{a}$ is the column vector $\underset{\sim}{a} = \left\{ a_1, \ldots, a_N \right\}$.

$\underset{\sim}{A}$, $\underset{\sim}{B}$ are N x N matrices having the elements

$$\left.\begin{aligned} A_{jk} &= \left(V_j^{IV} + 2pV_j'', V_k\right) \\ B_{jk} &= \left(V_j', V_k\right) \end{aligned}\right\} \quad \text{cf. sect. 7.4} \qquad (G)$$

In general, $A_{jk} \neq A_{kj}$, $B_{jk} \neq B_{kj}$.

$\underset{\sim}{I}$ is the identity matrix (the V_n are orthogonal and assumed to be normalized).

The approximate expression, (E), satisfies the boundary conditions (D) since the boundary conditions are the same for the systems shown in Fig. 13.14 and Fig. 13.15. (If for a given problem no functions, V_n, can be found which satisfy all boundary conditions, Galerkin's method may be replaced by the expressions of the principle of virtual work, cf. sect. 9. Then the geometrical boundary conditions only have to be satisfied).

(F) constitutes a matrix eigenvalue problem which has, say, the eigenvalues γ_j ; and the eigenvectors $\underset{\sim}{a}_j$ (see R. Zurmühl, Matrizen, Springer, Berlin).

Introducing $\underset{\sim}{a}_j$ into (E) we obtain an approximate eigenfunction $U_j(\xi)$ for (C), (D) which is correlated to the approximate eigenvalue γ_j.

H. Leipholz (Stabilitätstheorie, Teubner, Stuttgart 1968) investigated the convergence of this procedure for some classes of nonconservative problems. Furthermore, he discussed the applicability of the static stability criterion depending on the structure of the matrices $\underset{\sim}{A}$ and $\underset{\sim}{B}$ in equation (F).

Lit. see survey article : G. Herrmann, Stability of equilibrium of elastic systems subjected to nonconservative forces, Appl. Mech. Reviews, 20, No. 2 (1967).

14. Kinetic stability of non-autonomous systems

14. 1 A simple example (compressed bar)

Incompressible column (Euler-Bernoulli beam), see Fig. 14. 1, loaded axially by a periodic force $P(t) = P_0 + P_1 \cos \Omega t$

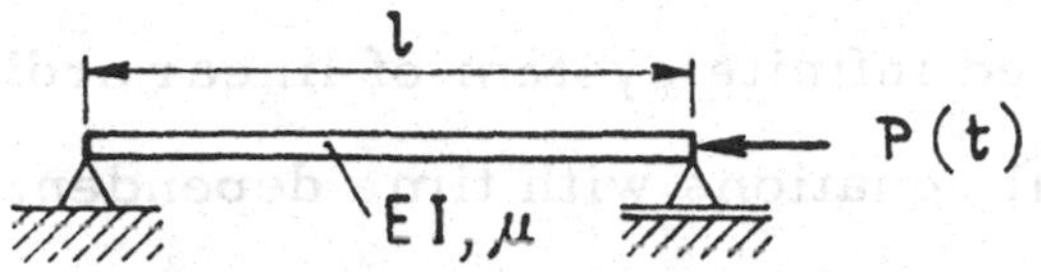

Fig. 14. 1

Equation of motion :

$$u^{IV} + p(\tau) u'' + \ddot{u} = 0 \qquad \text{(A)}$$

where $p(\tau) = P\left(\tau / \sqrt{EI/\mu l^4}\right) l^2 / EI = p_0 + p_1 \cos \omega t$,

$$\omega = \Omega / \sqrt{EI/\mu l^4}, \text{ cf. sect. 13. 1}$$

Boundary conditions :

$$u(0,\tau) = u''(0,\tau) = u(1,\tau) = u''(1,\tau) = 0 \qquad \text{(B)}$$

Assuming solutions of the form

$$u_n = a_n(\tau) \sin n\pi\xi \quad , \quad a_n\text{-normal coordinates,} \qquad \text{(C)}$$

we satisfy the boundary conditions, (B), and obtain from (A)

$$\ddot{a}_n + n^4\pi^4 \left[1 - \frac{p(\tau)}{n^2\pi^2}\right] a_n = 0 , \quad n = 1, \ldots \tag{D}$$

(D) is an uncoupled infinite system of linear ordinary second order differential equations with time dependent coefficients. These equations are of Hill's type if $p(\tau)$ is periodic, they are Mathieu-equations if $p(\tau) = p_0 + p_1 \sin \omega t$ as in our example.

14.11 Some results on Mathieu equations

(See, e.g., J. Meixner, F. W. Schäfke, Mathieusche und Sphäroidfunktionen..., Springer, Berlin 1954; Bolotin's book cited below).

We write (D) in the form

$$\frac{d^2y}{dx^2} + (\lambda - h^2 \cos 2x)\, y = 0 \tag{E}$$

By Floquet's theorem (E) has two linearly independent solutions of the form

$$y_F = q(x)\, e^{\nu x} \tag{F}$$

(there is an exceptional case). $q(x)$ is a (complex) periodic function

$$q(x+\pi) = q(x) .$$

ν - characteristic exponent (complex!).

The Floquet solutions , y_F , increase exponentially with x (the trivial solution $y \equiv 0$ of (F) is unstable) if $\mathrm{Re}\,\nu > 0$, they do not increase, we have (weak) stability, if $\mathrm{Re}\,\nu \leq 0$. ν depends on λ and h^2 , the parameters of (F). Fig. 14.2 shows stable and unstable parameter regions.(The chart is symmetrical with respect to the λ-axis).

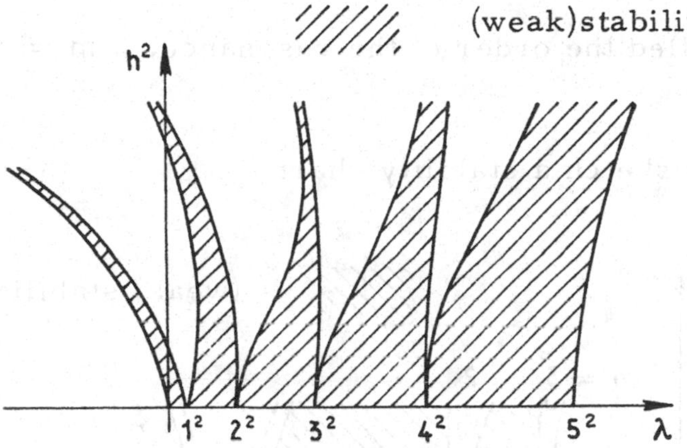

Fig. 14.2

On the calculation of such stability charts see the references cited in the literature mentioned above.

14. 12. Stability discussion for the given problem

Comparing (D) with (E) we obtain

$$\lambda = \left(\frac{2\,\omega_n}{\omega}\right)^2 , \qquad h^2 = \frac{4\,\omega_n\,p_1}{\omega^2}$$

where $\omega_n^2 = n^4\pi^4 - n^2\pi^2 p_0$ is the reduced natural frequency of the statically compressed beam. (Assumption $p_0 \ll p_{crit,\,stat.}$).

From Fig. 14.2 we see that, for small p_1, we have unstable regions ("resonance") in the vicinity of

$$\frac{2\,\omega_n}{\omega} = m \, , \qquad m = 1, 2, 3 \ldots \, . \tag{G}$$

(m is called the order of the resonance, $m = 1$ -main resonance).

We sketch a stability chart :

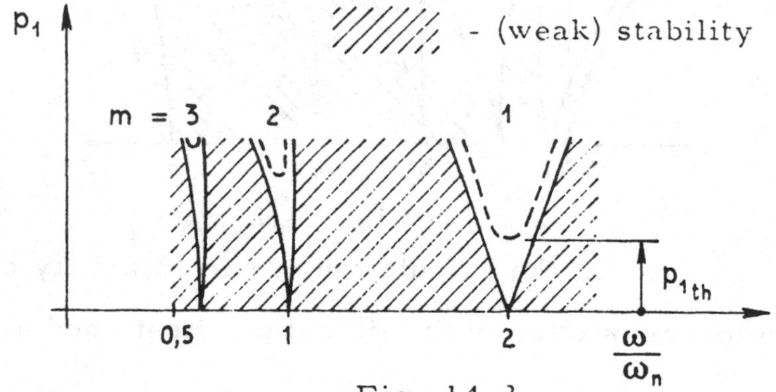

Fig. 14.3

Damping has two effects :

The weakly stable regions become asymptotically stable.

2. The stable regions grow, threshold values, p_{1th}) come into existence, i. e., the instability regions in Fig. 14. 3 are elevated.

Cf. sect. 62 (by E. Mettler) in W. Flügge's Handbook cited in sect. 2. 24

14. 2 A non-separable problem (compressed bar)

In section 14. 1 we were able to separate the variables, cf. equation (C). Let us consider now an example where no separation is possible : The column built in at both its ends.

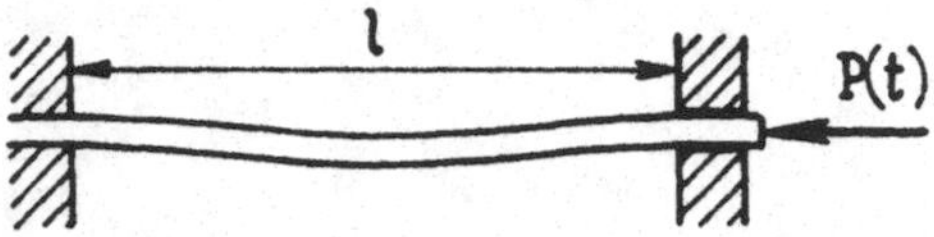

Fig. 14. 4

The equation of motion, $P(t)$ etc., is the same as in section 14. 1. The boundary conditions are

$$u(0,\tau) = u'(0,\tau) = u(1,\tau) = u'(1,\tau) = 0$$

Again we apply the eigenfunctions, V_n (cf. 13. 3 and 14. 1) , of the corresponding conservative system, i. e., the clamped-clamped beam compressed by p_0 , to express $u(\xi,\tau)$. But here, different from 14. 1, the systems of the functions V_n , V_n'' , V_n^{IV} are not simultaneously orthogonal. Thus we can not obtain a decoupled system of ordinary differential equations for the normal coordinates a_n , cf. equation (D) in the preceding section.

Introducing

$$u(\xi,\tau) = \sum_{n=1}^{N} a_n(\tau)\, V_n(\xi) \qquad \text{(A)}$$

which satisfies the boundary conditions, into the equations of motion we obtain by Galerkin's method

$$\underset{\sim}{A}\, \ddot{\underset{\sim}{a}} + \underset{\sim}{B}\, \underset{\sim}{a} + p_1 \cos\omega\tau\, \underset{\sim}{C}\, \underset{\sim}{a} = 0\,. \qquad \text{(B)}$$

$\underset{\sim}{a} = \left\{ a_1, \ldots, a_N \right\}$ is the column vector of the a_n, $\underset{\sim}{A}$, $\underset{\sim}{B}$, $\underset{\sim}{C}$ are $N \times N$ matrices. $\underset{\sim}{A}$ and $\underset{\sim}{B}$ are diagonal matrices, $\underset{\sim}{C}$ is a symmetric matrix, $C_{jk} = C_{kj}$ (cf. F. Weidenhammer, Der eingespannte, achsial pulsierend belastete Stab als Stabilitätsproblem, Ing.-Arch. 19 (1951), 162-191).

Equation (B) represents a finite system of linear ordinary differential equations with periodic coefficients. Again Floquet's theorem is valid (for first order systems see Coddington, Levinson's book cited in sect. 5.2, special results for second order systems see Bolotin's book cited below). Therefore, (B) has, except for some special cases, N linearly independent Floquet solutions

$$\underset{\sim}{a}_F(\tau) = \underset{\sim}{q}(\tau)\, e^{\nu\tau}, \qquad \text{(C)}$$

where

$$\underset{\sim}{q}\left(\tau + \frac{2\pi}{\omega}\right) = \underset{\sim}{q}(\tau) \qquad \text{(D)}$$

is periodic and ν is the characteristic exponent as in section 14.11.

The stability criteria are the same as in section 14.11. The calculation of stability regions is very troublesome and intricate. For small p_1 perturbation techniques are suitable (see, e.g., E. Mettler, Allgemeine Theorie der Stabilität erzwungener Schwingungen elastischer Körper, Ing. Arch. 17 (1949), 418-449). For special, numerically given, larger parameters the stability can be investigated by a numerical integration of the differential equation (B).

14.21 Qualitative discussion of some results

Mettler showed, cf. the paper cited in the preceding section, that in the general case instability regions start not only from the points $\omega = 2\omega_n / m$ of the ω-axis, cf. section 14.12, but also from the points

$$\omega = \frac{\omega_n + \omega_k}{m} \tag{E}$$

ω_n, ω_k - natural frequencies of the (conservative) autonomous system.

The instability regions starting from $\omega = (\omega_n + \omega_k)/m$, $\omega_n \neq \omega_k$, are called instability regions of the second kind. (The instability regions of the first kind and those of the second kind coincide for the example 14. 1). For the beam shown in Fig. 14. 4 results can be found in F. Weidenhammer, Der eingespannte, achsial pulsierend belastete Stab als Stabilitätsproblem, Ing. -Arch. 19(1951), 162-191.

The effect of damping terms depends on the assumptions about the magnitude of the damping coefficients. If the damping coefficients are introduced proportional to p_1, cf. equation (B), then damping may increase the angle of the wedge-shaped region of instability, cf. Fig. 14. 5. If, however the damping coefficients are independent of p_1, then a threshold value p_{1th}, similar as in Fig. 14. 3, appears but the width of the instability region can be increased, cf. Fig. 14. 6 (G. Schmidt, F. Weidenhammer, Instabilitäten gedämpfter rheolinearer Schwingungen, Math. Nachr. 23 (1961), 301).

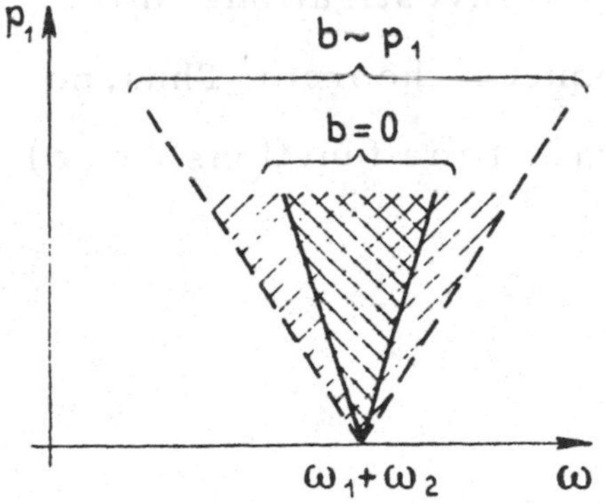

Fig. 14. 5

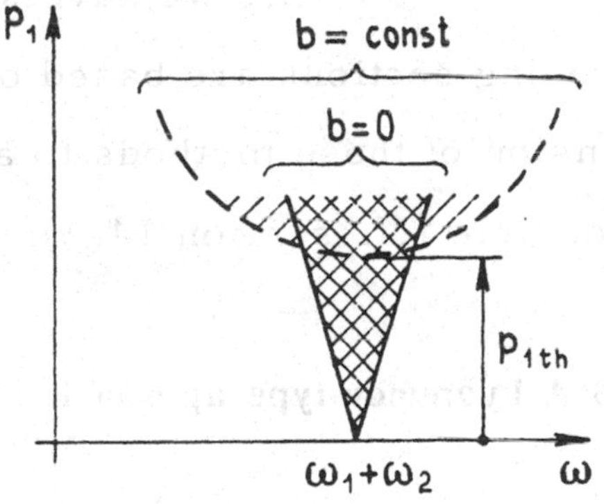

Fig. 14. 6

14.22 Two remarks

If solutions of the form (A) converge for $N \to \infty$, then there exist Floquet solutions of the linear periodic partial differential equation which describes the motion of this system (cf. equation (A) in sect. 14.1),

$$u(\xi,\tau) = Q_F(\xi,\tau)\, e^{\nu\tau} \tag{F}$$

where

$$Q_F(\xi,\tau + 2\pi/\omega) = Q(\xi,\tau) \tag{G}$$

To my knowledge for partial differential equations a Floquet theorem is still not proved.

As we have seen, the investigations in the preceding sections are based on Floquet's theorem. Thus, no extension of these methods to arbitrary time functions $p(\tau)$ is possible (cf. section 14.3).

14.3 A Lyapunov-type approach

We consider some results recently published: J.R. Dickerson, T.K. Caughey, Stability of continuous dynamic systems with parametric excitation, J. Appl. Mech.

36 E (1969), 212-216. (We sketch here a simplified "linearized" version of the theorem A of that paper).

For the system shown in Fig. 14.4 we have the equation of motion (cf. section 14.1)

$$\ddot{u} + b\dot{u} + u^{IV} + p(\tau)u'' = 0 \tag{A}$$

and the boundary conditions

$$u(0,\tau) = u'(0,\tau) = u(1,\tau) = u'(1,\tau) = 0 \,. \tag{B}$$

Damping included, $b > 0$ - damping coefficient.

(A) is a special case of the general equation (we call the variables t and X again)

$$\ddot{u} + L_1\dot{u} + L_2 u + L_3(t)u = 0 \,. \tag{C}$$

L_1, L_2 are constant (time independent!) differential operators with respect to X or coefficients (they might depend on X), $L_3(t)$ may depend on the time, too.
(In (A) : $L_1 = b$, $L_2 = \partial^4/\partial X^4$, $L_3 = p(t)\,\partial^2/\partial X^2$) .

We introduce an inner product :

$$(u,v) = (v,u) = \int_0^1 u(X,t)\,v(X,t)\,dX; \quad u,v \text{ - real}, \tag{D}$$

and define a Lyapunov function

$$V(u) = (L_1 u, L_1 u)/4 + (u, L_2 u) + ((L_1 u/2 + \dot{u}), (L_1 u/2 + \dot{u})) . \tag{E}$$

We assume

$$(u, L_2 u) \geq 0 , \tag{A1}$$

which is easily verified for our example. Because of (A1) the Lyapunov function $V(u)$ is positive definite :

$$V(u) \to 0 \hookrightarrow (L_1 u, L_1 u) \to 0 , (u, L_2 u) \to 0 . \tag{F}$$

Furthermore we assume

$$(\dot{u}, L_2 u) = (u, L_2 \dot{u}) \tag{A2}$$

which, again, is true for (A).

From (E) we obtain by differentiation with respect to t and elimination of $\ddot{u}$ by means of (C), taking (A2) into account,

$$\frac{dV}{dt} = -[(\dot{u}, L_1 \dot{u}) + (L_1 u, L_2 u) + (L_1 u, L_3 u) + 2(\dot{u}, L_3 u)] . \tag{G}$$

We demand

$$(u, L_1 u) \geq \lambda (u, u) , \quad \lambda > 0 , \tag{A3}$$

(is satisfied for (A) if $b > 0$) apply this inequality to the right hand side of (G) and put those terms into a suitable order :

$$\frac{dV}{dt} \leq - \lambda\varepsilon \left(\left(\dot{u} + \frac{L_3 u}{\lambda\varepsilon} \right) , \left(\dot{u} + \frac{L_3 u}{\lambda\varepsilon} \right) \right) -$$

$$- \left[(1-\varepsilon)\lambda(\dot{u}, \dot{u}) + (L_1 u , L_2 u) \right] +$$

$$+ \left[(L_3 u , L_3 u)/\lambda\varepsilon + |(L_1 u , L_3 u)| \right] . \tag{H}$$

The terms with ε are introduced to have some flexibility for some estimates which follow below.

dV / dt shall be estimated by a relation of the form

$$\frac{dV}{dt} \leq - \zeta V + \eta(t) V , \tag{I}$$

$\zeta , \eta(t)$ - positive. (I) is integrable :

$$V(t) \leq V_0 \exp \left\{ - \zeta t + \int_0^t \eta(\tau) \, d\tau \right\} . \tag{J}$$

To convert (H) into the form (I) we drop the first line on the right hand side of (H), (since that term is negative , the inequality is not violated). The second line and the third line of (H) are estimated in the following way :

$$M_1 \left[(1-\varepsilon)\lambda(\dot{u},\dot{u}) + (L_1 u, L_2 u) \right] \geq V \tag{K}$$

and

$$M_2 V \overset{\text{cf. (E)}}{\geq} M_2 \left[(L_1 u, L_1 u)/4 + (u, L_2 u) \right] \geq$$

$$\left[(L_3 u, L_3 u)/\lambda\varepsilon + |(L_1 u, L_3 u)| \right] . \tag{L}$$

To simplify M_1 in (K) we estimate V by

$$\left[3(L_1 u, L_1 u)/4 + (u, L_2 u) + 2(\dot{u},\dot{u}) \right] \geq V$$

which is obtained from (E) by using

$$\left((L_1 u/2 + \dot{u}), (L_1 u/2 + \dot{u}) \right) \leq (L_1 u, L_1 u)/2 + 2(\dot{u},\dot{u})$$

Thus ,we obtain from (K) and (L)

$$M_1(\varepsilon) \left[(1-\varepsilon)\lambda(\dot{u},\dot{u}) + (L_1 u, L_2 u) \right] \geq$$

$$\left[3(L_1 u, L_1 u)/4 + (u, L_2 u) + 2(\dot{u},\dot{u}) \right] \tag{A4}$$

and

$$M_2(\varepsilon,t)\left[(L_1u, L_1u)/4 + (u, L_2u)\right] \geq$$

$$\left[(L_3u, L_3u)/\lambda\varepsilon + |(L_1u, L_3u)|\right]. \qquad \text{(A5)}$$

(A4) and (A5) can be read as definitions (or conditions) for M_1 and M_2, respectively.

Now, the second and the third line of (H) are limited by $-V/M_1$ and VM_2, respectively. We have, cf. (H) and (I),

$$\zeta = \frac{1}{M_1(\varepsilon)}, \qquad \eta = M_2(\varepsilon,t) \qquad \text{(M)}$$

From (J) we obtain

$$V \leq V_0 \exp\left\{-\frac{t}{M_1(\varepsilon)} + \int_0^t M_2(\varepsilon,\tau)\,d\tau\right\} \leq V_0 e^{-\alpha t}, \qquad \text{(N)}$$

where

$$\alpha = \frac{1}{M_1} - \sup_t \frac{1}{t}\int_0^t M_2(\varepsilon,\tau)\,d\tau \qquad \text{(O)}$$

The trivial solution, $u \equiv 0$, of (C) is asymptotically stable if the assumptions (A1)-(A5) are satisfied and if

$\alpha > 0$. Because of (F) we have

$$(L_1 u, L_2 u)/4 \, , \; (u, L_2 u) < V_0 e^{-\alpha t} \tag{P}$$

We apply (N), (O) to the example (A) :

The assumptions (A1)-(A3) are satisfied. (A4) has the form

$$M_1 \left[(1-\varepsilon) b (\dot{u}, \dot{u}) + (bu, u^{IV}) \right] \geq \left[3 (bu, bu)/4 + (u, u^{IV}) + 2 (\dot{u}, \dot{u}) \right] \tag{Q}$$

(Q) is valid if

$$M_1 (1-\varepsilon) b \geq 2 \tag{R}$$

and

$$M_1 b (u, u^{IV}) \geq 3 b^2 (u, u)/4 + (u, u^{IV}) \tag{S}$$

To estimate (u,u) by $(u, u^{IV}) = (u'', u'')$ we start from

$$u = \int_0^x u' \, d\xi \quad , \qquad u' = \int_0^x u'' \, d\xi \; ,$$

cf. the boundary conditions (B), and obtain by Schwarz's inequality

$$u^2 \leq X \int_0^X (u')^2 \, d\xi \quad , \quad u'^2 \leq X \int_0^X (u'')^2 \, d\xi$$

and

$$(u,u) \leq (u'',u'') = (u, u^{IV}) \, .$$

Thus,(R) and (S) are satisfied by

$$M_1 = \max \left[\frac{2}{(1-\varepsilon) b} \, , \, \frac{3b^2/4 + 1}{b} \right] .$$

From (A5) we obtain, $p = p(t)$,

$$M_2 \left[(bu, bu)/4 + (u, u^{IV}) \right] \geq \left[(pu, pu)/\lambda\varepsilon + |(bu, pu)| \right] . \tag{T}$$

(T) is satisfied by

$$M_2 = \left[\frac{p^2(t)}{b\varepsilon} + b\,|p(t)| \right]$$

Therefore,we have asymptotic stability if

$$\sup \frac{1}{t} \int_0^t \left[\frac{p^2(\tau)}{b\varepsilon} + b\,|p(\tau)| \right] d\tau < \frac{b}{\max\left[\frac{2}{1-\varepsilon} \, , \, 3b^2/4 + 1\right]} \, . \tag{U}$$

For $| p(\tau) | > 1$ no stability is guaranteed by (U). The static critical load is $p_{crit, stat} = 4\pi^2$.

14. 4 Concluding remarks

We have restricted ourselves to linear investigations. Some nonlinear effects will be discussed later.

There exist many investigations "on the parametric response of structures". In a survey article of that title by R. M. Evan-Iwanowski (Appl. Mech. Reviews 18, No. 9 (1965)) some literature can be found. Special prominence has to be given to the book : V. V. Bolotin, The dynamic stability of elastic systems, Holden-Day, San Francisco 1964. A series of interesting papers is contained in the proceedings of a conference held in 1965 : G. Herrmann (ed.), Dynamic stability of structures, Pergamon Press, Oxford 1967.

15. Nonlinear vibrations of autonomous systems

15. 1 Nonlinear torsional vibrations of a cylinder

(cf. H. Kauderer, Nichtlineare Mechanik, Springer, Berlin 1958).

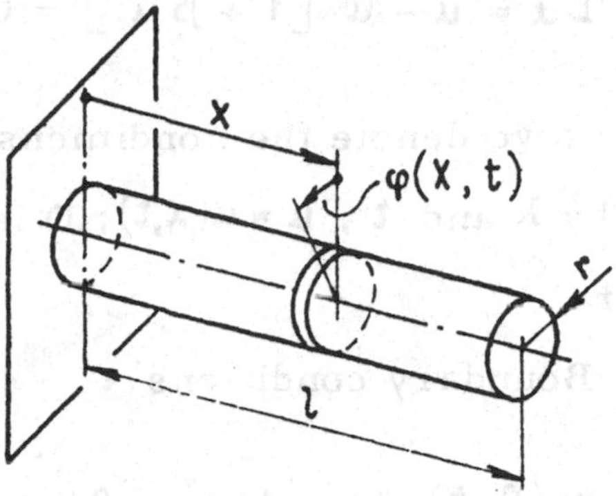

Fig. 15.1

φ - angle of twist

M_T - torque

$I_p = \pi r^4/2$ - polar moment of inertia

ϱ - density

G - shear modulus

Acceleration of an element

$$\varrho I_p \frac{\partial \varphi}{\partial t^2} = \frac{\partial M_T}{\partial x}$$

Nonlinear (cubic)stress-strain relation leads to

$$M = G I_p \frac{\partial \varphi}{\partial x}\left[1 + \alpha\left(\frac{\partial \varphi}{\partial x}\right)\right] \;, \quad \alpha - \text{ parameter}$$

Elimination of M_T and introduction of non-dimensional quantities yield the equation of motion

$$Du = \ddot{u} - u''\left[1 + \beta u'^2\right] = 0. \qquad (*)$$

$u \sim \varphi$, $\beta \sim \alpha$; we denote the nondimensional independent variables again by X and t , $u = u(X,t)$; D is a non linear differential operator.

Boundary conditions

$$u(0,t) = u'(1,t) = 0 . \qquad (**)$$

The equation of motion if nonlinear, the boundary conditions are linear, no superposition is possible. The system is conservative

15.11 Solution of the linear problem

For $\beta = 0$ we have a linear problem. Its solutions (eigenfunctions, eigenvalues, cf. sect. 5) are

$$u_n = U_n(X)\, T_n(t) = \sin \zeta_n X \left(A_n \cos \omega_n t + B_n \sin \omega_n t\right) . \qquad (A)$$

$$\zeta_n = \pi (n - 1/2) \, , \quad \omega_n = \zeta_n \quad , \quad n = 1, 2, \dots \qquad (B)$$

The system of the eigenfunctions, U_n, is complete.

15.12 Reduction of (*) to a system of ordinary differential equations

We assume that (*) has an approximate solution of the form

$$u_A(X,t) = \sum_{n=1}^{N} a_n(t)\,U_n(X). \tag{C}$$

Since the U_n are the eigenfunctions of the linear problem, (C) satisfies the boundary conditions (**) for arbitrary $a_n(t)$.

By Galerkin's method we obtain

$$0 = (Du_A, U_n) = \ddot{a}_n + \omega_n^2 a_n + \beta f_n(a_1,\dots,a_N),\ n = 1,\dots,N. \tag{D}$$

The $f_n(a_1,\dots,a_N)$ are cubic functions of their arguments. We abbreviate (D) in matrix notation by

$$\ddot{\underset{\sim}{a}} + \underset{\sim}{C}\,\underset{\sim}{a} + \beta \underset{\sim}{f}(\underset{\sim}{a}) = 0, \tag{E}$$

where $\underset{\sim}{C} = \mathrm{diag}(\omega_n^2),\ n = 1,\dots,N.$

For $N = 2$ we have

$$\left.\begin{aligned} \ddot{a}_1 + \omega_1^2 a_1 + \beta^*\left[a_1^3 + 3\, a_1^2\, a_2 + 18\, a_1\, a_2^2\right] = 0 \\ \ddot{a}_2 + \omega_2^2 a_2 + \beta^*\left[a_1^3 + 18\, a_1^2\, a_2 + 81\, a_2^3\right] = 0 \end{aligned}\right\} \quad (F)$$

$\beta^* = \beta \pi^4 / 64$ (below we write β instead of β^*).

15. 2 First order approximation $N = 1$

For $N = 1$ we obtain from equation(F) of the preceding section ($a_1 = a$):

$$E a \equiv \ddot{a} + \omega_1^2 a + \beta a^3 = 0 . \qquad (A)$$

This is Duffing's equation. E-nonlinear operator.

The solutions of equations of the type

$$\ddot{a} + h(a) = 0 \,, \quad a\, h(a) > 0 \quad \text{for} \quad a \neq 0 \,, \ |a| \leq |a|_{max} \,, \qquad (B)$$

are periodic and can always be reduced to an integration (because the conservation of energy represents a first integral of (B)). For Duffing's equation, (A), we obtain

$$t - t_0 = \int_{a(t_0)}^{a(t)} \frac{da}{\sqrt{\dot{a}^2(t_0) - \omega_1^2 a^2 - \beta a^4/2}}$$

a (t) can be expressed by elliptic functions (cf. Kauderer's book cited above).

In general, it is impossible to find analytic solutions for higher order approximations, $N > 1$.

15.21. One term Fourier approximation

As a rough approximate solution of (A) we choose

$$a(t) = A \cos \omega t . \qquad \text{(C)}$$

Putting (C) into (A) and expanding $E\left[A \sin \omega t\right]$ into a Fourier series we get

$$E\left[A \sin \omega t\right] = \underbrace{\left[-A\omega^2 + A\omega_1^2 + 3\beta A^3/4\right]}_{\stackrel{!}{=}\, 0} \cos \omega t + \underbrace{A^3/4 \cos 3\omega t}_{\text{omitted}} \qquad \text{(D)}$$

We obtain

$$\omega^2 = \omega_1^2 + 3\beta A^2/4 \qquad \text{(E)}$$

see Fig. 15.2 ($\beta > 0$: strain hardening; $\beta < 0$: strain softening).

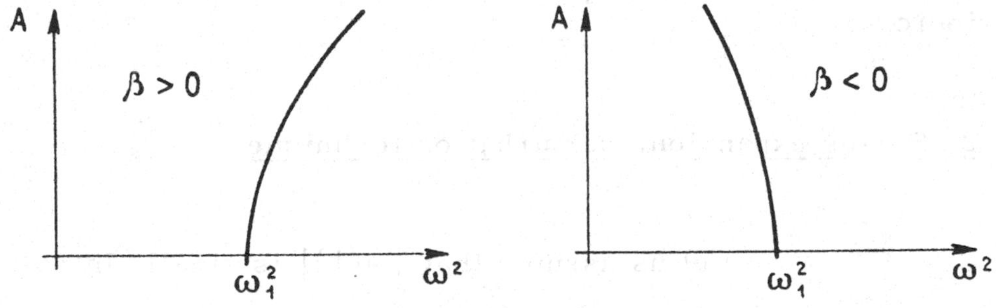

Fig. 15.2

The frequency, ω, depends on the amplitude, A, of the oscillation.

Similar simple assumptions, $a_n = A_n \cos \omega t$, can be introduced into higher order approximations, $N > 1$, cf. equations (D) and (F) in section 15.12. But, in general, it may be necessary to introduce solutions of the form

$$a_n = A_n \cos \omega t + B_n \sin \omega t .$$

Since the system is autonomous, the origin of the time axis can be shifted in such a way that, say, B_1 vanishes. By a procedure similar to (D) we obtain $2N$ (nonlinear) equations for the $2N$ unknowns $A_1, A_2, B_2, \ldots, A_N, B_N$ and ω, cf. section 15.4.

If the system of ordinary differential equations (D) in section 15.12 has solutions of this type, ω and the coefficients A_j, B_j are functions of β and of one of the coefficients A_{j*}, say ; u_A, cf. (C) in sect. 15.12, is periodic.

15.22 Series expansion, perturbation technique

Let us assume that $|a(t)|$ is small. Introduction of

$$a = \varepsilon\bar{a} \quad , \quad \varepsilon\text{-small parameter}$$

into (A) yields

$$\ddot{\bar{a}} + \omega_1^2 \bar{a} + \underbrace{\varepsilon^2\beta}_{\text{small}} \bar{a}^3 = 0 . \qquad (F)$$

(F) corresponds to equation (A) with β small. So we have to investigate $a(t,\beta)$ for small values β .

We want to expand $a(t,\beta)$ with respect to β Looking at (C) and (E) we can expect to obtain terms of the following kind in that expansion :

$$\begin{aligned} a(t) &= A \cos\left(\omega_1 + \zeta(\beta)\right) t + \ldots \\ &= A \cos \omega_1 t + A\,\zeta(\beta) t \sin \omega_1 t + \ldots \end{aligned}$$

Even if this expansion would converge we would not be able to interprete the terms in the second line because of the secular terms, t , t^2 , ... Especially, we would not be able to determine the period of the solution

$$T(\beta) = \frac{2\pi}{\omega(\beta)} \quad ; \quad a(t+T,\beta) = a(t,\beta) \qquad (G)$$

To overcome this difficulty we introduce a new time scale (Lindstedt, 1883).

$$\tau = \omega t . \qquad (H)$$

$a(\tau,\beta)$ is 2π- periodic with respect to τ . From (A) we obtain

$$\omega^2 \frac{d^2 a}{d\tau^2} + \omega_1^2 a + \beta a^3 = 0 . \qquad \text{(I)}$$

(We denote $d^2 a / d\tau^2$ by $\ddot{a}$).

Now we expand a and ω^2 into series with respect to β

$$\left.\begin{aligned} a(\tau,\beta) &= a^{(0)} + \beta a^{(1)} + \beta^2 a^{(2)} + \ \ldots \\ \omega^2(\beta) &= \omega_1^2 + \beta \varkappa_1 + \beta^2 \varkappa_2 + \ \ldots \end{aligned}\right\} \qquad \text{(J)}$$

Putting (J) into (I) and comparing terms with equal powers of β we obtain

$$\left.\begin{aligned} &\omega_1^2 \ \ddot{a}^{(0)} + \omega_1^2 a^{(0)} = 0 \ , \quad \text{generating equation} \\ &\omega_1^2 \ \ddot{a}^{(1)} + \omega_1^2 a^{(1)} = -\left(a^{(0)}\right)^3 - \varkappa_1 \ddot{a}^{(0)} \\ &\omega_1^2 \ \ddot{a}^{(2)} + \omega_1^2 a^{(2)} = -3\left(a^{(0)}\right)^2 a^{(1)} - \varkappa_1 \ddot{a}^{(1)} - \varkappa_2 \ddot{a}^{(0)} \end{aligned}\right\} \qquad \text{(K)}$$

Initial conditions :

$$a(0) = A \ , \quad \dot{a}(0) = 0 \qquad \text{(L)}$$

(L) is satisfied if

$$\left.\begin{aligned} &a^{(0)}(0) = A \ , \quad a^{(j)}(0) = 0 \ , \quad j = 1,2,\ldots \\ &\dot{a}^{(j)}(0) = 0 \ , \quad j = 0,1,2,\ldots \end{aligned}\right\} \qquad (M)$$

Furthermore all $a^{(j)}(\tau)$ must be 2π-periodic.

From the first equation (K) we obtain

$a^{(0)} = A \cos \tau$, generating solution,

$a(\tau, \beta) \to a^{(0)}$ for $\beta \to 0$.

Putting $a^{(0)}$ in the second equation (K) we get

$$\omega_1^2 \ddot{a}^{(1)} + \omega_1^2 a^{(1)} = - A^3 \left[\frac{3}{4} \cos \tau + \frac{1}{4} \cos 3\tau \right] + \varkappa_1 A \cos \tau \qquad (N)$$

Since $a^{(1)}$ shall be 2π-periodic, we have to choose

$$\varkappa_1 = \frac{3}{4} A^2 . \qquad (O)$$

The 2π-periodic solution $a^{(1)}$ of (N) which satisfies the initial conditions given above is

$$a^{(1)} = \frac{A^3}{32\,\omega_1^2} \left[\cos 3\tau - \cos \tau \right] ;$$

and so on.

In the first approximation we have, cf. (J), (O),

$$\omega^2 = \omega_1^2 + 3\beta A^2/4 . \qquad \text{(P)}$$

This expression coincides with equation (E) which was found by the one term Fourier approximation.

The preceding calculations represent a rather spacial outline of a general technique. Mathematical investigations of these procedures were originated by Poincaré. General theory, see: I. G. Malkin,

Some problems in the theory of nonlinear oscillations, transl. from Russian, AEC-tr-3766 (book 1/2), US Department of Commerce, Institute for Applied Technology.

In general a better convergence is achieved if (I) is devided by ω^2 ,

$$\ddot{a} + \frac{\omega_1^2}{\omega^2} a + \frac{1}{\omega^2} \beta a^3 = 0 \quad ,$$

and $1/\omega^2$ is expanded into a series,

$$\frac{1}{\omega^2} = \frac{1}{\omega_1^2} + \mu_1 \beta + \mu_2 \beta^2 + \dots .$$

See Malkin's book cited above and Yu. A. Ryabov, pp. 425 - 445 in vol. 1 of the Proceedings of the International Symposion on Nonlinear Vibrations, Published by the Academy of Sciences of the USSR, Kiev 1963.

If the nonlinear terms are continuously differentiable but not analytic, a series expansion is impossible but a successive approximation technique yields results, cf. Malkin's book.

15. 3 Higher order approximations N > 1 .

15. 31 Periodic solutions; perturbation technique.

Example (F) from section 15. 12 :

$$\left.\begin{aligned} \omega^2\ddot{a}_1 + \omega_1^2 a_1 + \beta\left[a_1^3 + 3a_1^2 a_2 + 18 a_1 a_2^2\right] &= 0 \\ \omega^2\ddot{a}_2 + \omega_2^2 a_2 + \beta\left[a_1^3 + 18 a_1^2 a_2 + 81 a_2^3\right] &= 0 \end{aligned}\right\} \quad \text{(A)}$$

Here we have already transformed t into τ :

$$\omega t = \tau \qquad \text{(cf. (H) in sect. 15. 22).}$$

Let us look for 2π-periodic functions $a_1(\tau, \beta)$, $a_2(\tau, \beta)$ and an $\omega(\beta)$ which for $\beta \to 0$ tend towards

$$\left.\begin{aligned} a_1(\tau, \beta) &\to A \cos \tau \\ a_2(\tau, \beta) &\to 0 \\ \omega(\beta) &\to \omega_1 \end{aligned}\right\} \quad \text{(B)}$$

(That corresponds to the first normal mode, cf. (A), (C) in sect. 15.11).

Introducing

$$\left.\begin{aligned} \omega^2 &= \omega_1^2 + \beta \varkappa_1 + \beta^2 \varkappa_2 + \dots \\ a_1 &= a_1^{(0)} + \beta a_1^{(1)} + \beta^2 a_1^{(2)} + \dots \\ a_2 &= a_2^{(0)} + \beta a_2^{(1)} + \beta^2 a_2^{(2)} + \dots \end{aligned}\right\} \quad (C)$$

into (A) we obtain

$$\left.\begin{aligned} \omega_1^2 \ddot{a}_1^{(0)} + \omega_1^2 a_1^{(0)} &= 0 \\ \omega_1^2 \ddot{a}_2^{(0)} + 9\omega_1^2 a_2^{(0)} &= 0 \end{aligned}\right\} \quad \text{generating equations (D)}$$

(Note that $\omega_2 = 3\,\omega_1$).

Because of (B) the (tentative) generating solutions of (D) are

$$\left.\begin{aligned} a_1^{(0)} &= A \cos \tau \\ a_2^{(0)} &\equiv 0 \end{aligned}\right\} \quad (E)$$

(All solutions $A_2 \cos 3\tau + B_2 \sin 3\tau$ of the second equation (D) are 2π-periodic. By the restriction (B) we eliminated this arbitrariness).

Taking (E) into account we get from (A) and (C) the following equations for $a_1^{(1)}$, $a_2^{(1)}$

$$\omega_1^2 \ddot{a}^{(1)} + \omega_1^2 a_1^{(1)} = -A^3\left[\frac{3}{4}\cos\tau + \frac{1}{4}\cos 3\tau\right] + \varkappa_1 A\cos\tau \qquad (F)$$

$$\omega_1^2 \ddot{a}_2^{(1)} + 9\,\omega_1^2 a_2^{(1)} = -A^3\left[\frac{3}{4}\cos\tau + \frac{1}{4}\cos 3\tau\right] \qquad (G)$$

In (F) resonance can be avoided by choosing

$$\varkappa_1 = \frac{3}{4}A^2 .$$

In (G) internal resonance can not be prevented. The solutions (E), and the assumptions (B), do not lead to practicable equations.

We replace (E) by

$$\left.\begin{aligned} a_1^{(0)} &= A\cos\tau \quad , \ A = A_1 , \\ a_2^{(0)} &= A_2\cos 3\tau \end{aligned}\right\} \qquad (H)$$

and obtain for $a_1^{(1)}$, $a_2^{(1)}$

$$\omega_1^2 \ddot{a}_1^{(1)} + \omega_1^2 a_1^{(1)} = -\left[\left(a_1^{(0)}\right)^3 + 3\left(a_1^{(0)}\right)^2 a_2^{(0)} + 18\, a_1^{(0)}\left(a_2^{(0)}\right)^2\right] - \varkappa_1 \ddot{a}_1^{(0)}$$

$$\omega_1^2 \ddot{a}_2^{(1)} + 9\,\omega_1^2 a_2^{(1)} = -\left[\left(a_1^{(0)}\right)^3 + 18\left(a_1^{(0)}\right)^2 a_2^{(0)} + 81\left(a_2^{(0)}\right)^3\right] - \varkappa_1 \ddot{a}_2^{(0)}$$

By a proper choice of $A_2(A)$ and $\varkappa_1(A)$ resonance can be prevented in both equations.

General procedure see Malkin's book cited above.

We see :

1. In general, the generating solution can not be chosen arbitrarily.
2. A nonlinear periodic solution u_p, here approximated by u_A, does not, in general, branch off from a linear normal mode. In Fig. 15.3 such a behavior is shown qualitatively.

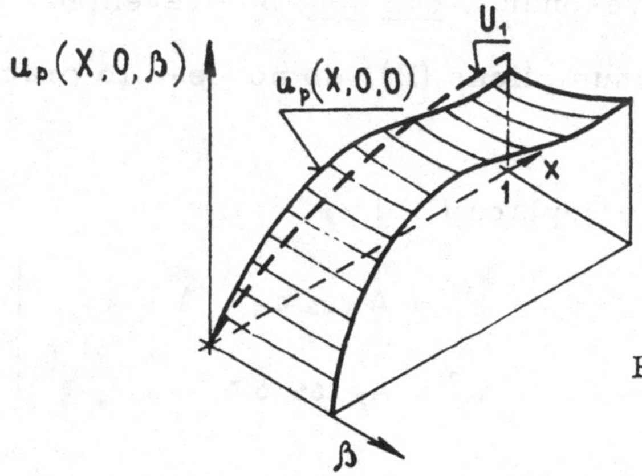

Fig. 15.3

"Reasonable" behavior occurs, (B) is possible etc., if the natural frequencies are not commensurable.

15.32 Non-periodic solutions; some remarks

Let us assume that we know initial conditions

$$\left.\begin{aligned} u(X,0) &= \varphi(X), \\ \dot{u}\; X,0 &= \psi(X) \end{aligned}\right\} \quad (I)$$

for the problem given in section 15.1. Similar as in section 5.5, we can derive initial conditions from (I) for the coordinate functions $a_n(t)$:

$$\left.\begin{aligned} a_n(0) &= a_{no} \\ \dot{a}_n(0) &= \dot{a}_{no} \end{aligned}\right\} \quad n = 1, \ldots, N. \qquad \text{(J)}$$

The equation (E) from section 15.12,

$$\ddot{\underset{\sim}{a}} + \underset{\sim}{C}\,\underset{\sim}{a} + \beta \underset{\sim}{f}(\underset{\sim}{a}) = 0\,, \qquad \text{(K)}$$

together with (J) establish an initial value problem for a system of ordinary differential equations.

Such a problem can be solved numerically for finite time intervals. If β is a small parameter, perturbation techniques can yield results (assuming that we are only interested in a finite time intervall even secular terms would not matter). Some special cases may be investigated by asymptotic methods (slowly varying phase and amplitude)which we shall discuss briefly in section 16; see N.N. Bogoliubov, Y.A. Mitropolsky, Asymptotic methods in the theory of nonlinear oscillations, Gordon & Breach, New York 1961; Y.A. Mitropolsky, B.I. Moseenkov, Lectures on the application of asymptotic methods to the solution of partial differential equations(in

Russian), Ukrainian Academy of Sciences, Kiev 1968.

15. 4 Numerical computation of periodic solutions

High order approximations for periodic solutions of nonlinear differential equations can be calculated only by numerical methods. M. Urabe and A. Reiter (Numerical computation of nonlinear forced oscillations by Galerkin's procedure, J. Math. Anal. Appl. 14 (1966) 107-140)describe a procedure which is applicable to ordinary differential equations. Here we shall extend this procedure to partial differentials equations.

Periodic solutions of (E),sect. 15. 12, can be found directly by Urabe's method (the restriction to forced vibrations in the paper cited above is not essential, see below). But in general it is very cumbersome to establish the set of equations (E). So lets start right from the original differential equation (*).

15. 41 Description of the method

First we introduce $\omega t = \tau$ into (*) and obtain

$$Du = \omega^2 \ddot{u} - u''\left[1 + \beta u'^2\right] = 0 . \qquad \text{(A)}$$

(After the time transformation we write t instead of τ, $\ddot{u} = \partial^2 u / \partial \tau^2$).

We are looking for a 2π-periodic solution of (A),

$$u(X, t + 2\pi) = u(X, t) . \qquad \text{(B)}$$

Again we start with

$$u(X,t) = \sum_{n=1}^{N} a_n(t)\, U_n(X) , \qquad \text{(C)}$$

cf. (C) in section 15. 12 (here we drop the subscript A). The $U_n(X)$ are the natural modes of the linear problem, cf. section 15. 11, or some other appropriate linearly independent functions which satisfy individually the boundary conditions(**).

β is a numerically given parameter.

Here the $a_n(t)$ are assumed to be 2π-periodic. We shall represent them by a truncated Fourier series. For simplicity let us assume that the $a_n(t)$ have the form

$$a_n(t) = \sum_{m=1}^{M} a_{nm} \cos(2m-1)t \quad , \; n = 1, \dots, N . \qquad \text{(D)}$$

We introduce

$$b_k = a_{nm} \ , \quad k = (n-1)M + m \tag{E}$$

and

$$u_k(X,t) = U_n(X)\cos(2m-1)t \ . \tag{F}$$

u , cf. (C),is represented by

$$u = \sum_{k=1}^{N \cdot M} b_k u_k = \mathcal{S}\,\underset{\sim}{b} \ , \tag{G}$$

where the symbol $\mathcal{S}$ stands for "Fourier synthesis" and $\underset{\sim}{b}$ is the vector

$$\underset{\sim}{b} = \{ b_1, \ldots, b_{NM} \} .$$

The operator $\mathcal{S}$ is linear,

$$\mathcal{S}(\underset{\sim}{b}_1 + \underset{\sim}{b}_2) = \mathcal{S}\underset{\sim}{b}_1 + \mathcal{S}\underset{\sim}{b}_2$$

The gradient of u with respect to $\underset{\sim}{b}$, denoted by u_b , has the form

$$u_b = \mathcal{S}\,\underset{\sim}{I} \ , \tag{H}$$

where $\underset{\sim}{I}$ denotes the $(NM) \times (NM)$ identity matrix.

We define the inner product

$$r_k = (u, u_k) = \int_0^1 \int_0^{2\pi} u \, u_k \, dX \, dt \ , \quad k = 1, \ldots, NM. \tag{I}$$

r_k - (generalized) Fourier coefficient.

By

$$\underset{\sim}{r} = \mathcal{A} u \tag{J}$$

we symbolize the generalized "Fourier analysis". The operator $\mathcal{A}$ is linear.

$$\mathcal{A}(u+v) = \mathcal{A}u + \mathcal{A}v . \tag{K}$$

In the forms given here $\mathcal{S}$ and $\mathcal{A}$ are not inverse operators,

$$\mathcal{A}\mathcal{S}\underset{\sim}{b} \neq \underset{\sim}{b} ,$$

because some factors in (I) are not chosen properly. (These factors depend on the U_n , a proper choice could be introduced but it is not essential for our investigations).

Let us assume that we know a set of numbers $\underset{\sim}{b}$ which represent an approximate solution u of Du . In general $D\mathcal{S}\underset{\sim}{b}$ will not vanish identically. We require

$$\underset{\sim}{r} = \mathcal{A} D \mathcal{S} \underset{\sim}{b} \overset{!}{=} 0 . \tag{L}$$

The residuum vector $\underset{\sim}{r}$ is a nonlinear function of $\underset{\sim}{b}$,

$$\underset{\sim}{r} = \underset{\sim}{r}(\underset{\sim}{b}) .$$

The vector equation

$$\underset{\sim}{r}(\underset{\sim}{b}) = 0 \tag{M}$$

represents a set of NM nonlinear scalar equations for the elements b_k, $k = 1,\ldots,NM$, of $\underset{\sim}{b}$. (We shall show how to solve the equations in the next section)

For numerically given values $\underset{\sim}{b}$ the residuum vector $\underset{\sim}{r}(\underset{\sim}{b})$ can be obtained by numerical methods; u and $\underset{\sim}{r}$ have to be computed corresponding to (G)and (I). (Attention : For the numerical integration the subdivisions of the X-interval and of the t-interval must be small enough. In general a fraction of $1/N$ and $2\pi/M$, say $1/3$, is necessary).

15.42 Solution of the equations

We apply Newton's method to improve the chosen $\underset{\sim}{b}$. Let $\Delta\underset{\sim}{b}$ denote the correction. From

$$\mathcal{A} D \mathcal{S}(\underset{\sim}{b} + \Delta\underset{\sim}{b}) = 0$$

we obtain

$$\mathcal{A} D(\mathcal{S}\underset{\sim}{b} + \mathcal{S}\Delta\underset{\sim}{b}) = 0 \qquad \text{(N)}$$

and by linearizing $D(\mathcal{S}\underset{\sim}{b} + \mathcal{S}\Delta\underset{\sim}{b})$ with respect to $\mathcal{S}\Delta\underset{\sim}{b}$ we get

$$\underbrace{\mathcal{A} D \mathcal{S}\underset{\sim}{b}}_{\underset{\sim}{r}(\underset{\sim}{b})} + \underbrace{(\mathcal{A} D_u \mathcal{S}\underset{\sim}{I})}_{\underset{\sim}{F}}\Delta\underset{\sim}{b} = 0 \qquad \text{(O)}$$

$$\underset{\sim}{r}(\underset{\sim}{b}) + \underset{\sim}{F}\Delta\underset{\sim}{b} = 0$$

$\underset{\sim}{r}(\underset{\sim}{b})$ is the residuum vector calculated for the chosen $\underset{\sim}{b}$;

D_u denotes the (Fréchet) derivative (linear variation) of D, for instance, in the case (*) :

$$D_u = \omega^2 \frac{\partial^2}{\partial t^2} - \left[1 + \beta (u')^2\right] \frac{\partial^2}{\partial x^2} - 2\beta u' u'' \frac{\partial}{\partial x} \, . \qquad \text{(P)}$$

$\underset{\sim}{F}$ is the matrix

$$\underset{\sim}{F} = \mathcal{A} D_u \mathcal{S} \underset{\sim}{I} \qquad \text{(Q)}$$

$\underset{\sim}{F}$ can be calculated numerically corresponding to equation (Q). In general, $\underset{\sim}{F}$ depends on $\underset{\sim}{b}$, cf. the terms $u = \mathcal{S}\underset{\sim}{b}$ in D_u , equation (P).

For $\det \underset{\sim}{F} \neq 0$ we obtain from (O)

$$\Delta \underset{\sim}{b} = - \underset{\sim}{F}^{-1} \underset{\sim}{b} \qquad \text{(R)}$$

and the improved Fourier coefficients

$$\underset{\sim}{b}^* = \underset{\sim}{b} + \Delta \underset{\sim}{b} \, .$$

Iterative application of this procedure leads (within the limits of numerical accuracy) to a solution $\underset{\sim}{b}^{**}$ of (L) if $\det \underset{\sim}{F} \neq 0$ in the neighbourhood of $u = \mathcal{S}\underset{\sim}{b}^{**}$ and the initial vector $\underset{\sim}{b}$ is chosen close enough to $\underset{\sim}{b}^{**}$.

There will be one difficulty we have to overcome. In general, the Fourier coefficients will change very rapidly when the frequency ω is changed (cf.

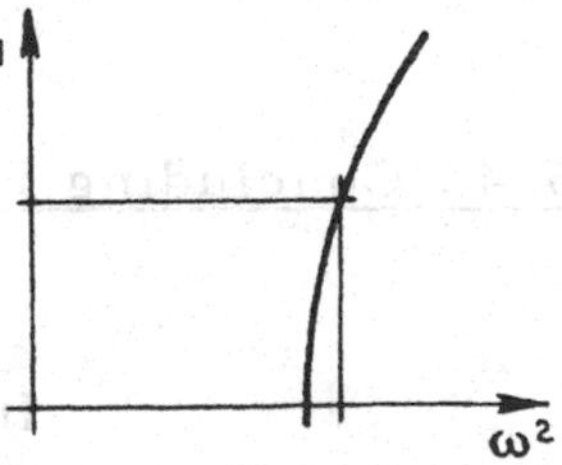

Fig. 15.4

Fig. 15.4 and Fig. 15.2. (ω is parameter in our calculations). Very frequently the iteration procedure outlined above will converge to $\underset{\sim}{b}^{**} = 0$, the trivial solution. This can be prevented by interchanging ω (or ω^2) and one of the elements b_k , say b_{k*} , in Newton's iteration process : b_{k*} is introduced as parameter and used to calculate $\underset{\sim}{r}(\underset{\sim}{b})$ as before, cf. equation (L). (A tentative value ω is estimated and introduced into that computation). For Newton's procedure we have $\Delta b_{k*} = 0$ and $\Delta\omega$ (or $\Delta\omega^2$) is put into the place of Δb_{k*} in $\Delta \underset{\sim}{b}$, $\underset{\sim}{F}$ is appropriately modified. By (R) we obtain the corrections $\Delta b_k \; (k \neq k^*)$ and $\Delta\omega$ (or $\Delta\omega^2$). Since $b_{k*} \neq 0$ this procedure can not converge to the trivial solution $\underset{\sim}{b} = 0$

In general,it is advisable to start the computation from parameters $\beta_0, (b_{k*})_0$, say, for which good approximate values b_k , ω can be guessed,and to apply $\underset{\sim}{b}_0^{**}$, ω_0^{**} , the numerically obtained solutions for β_0 , $(b_{k*})_0$, as starting approximations for β_1 , $(b_{k*})_1$ which differ slightly from $\beta_0, (b_{k*})_0$; etc.

15.43 Concluding remarks

In general,in equation (D) a more complete

Fourier series has to be taken into account. The computational work can be decreased considerably if the formal calculations included in $\mathfrak{D}_u \, \mathfrak{S} \, \underset{\sim}{I}$ are done beforehand analytically. Known special characteristics of the solutions should be taken into account.

The expression $\mathfrak{A} \, \mathfrak{D} \, \mathfrak{S} \, \underset{\sim}{b}$ in equation (L) can be replaced by the general form of Hamilton's principle, see equation (H) in section 9. (The δu would be the functions u_k, cf. equ. (F), for the example discussed above). Hamilton's principle has the advantage that nonlinear dynamic boundary conditions are included. Nonlinear geometric boundary conditions may be considered as constraints which can be treated, in connection with Hamilton's principle, by the method of Lagrangian multipliers. In that case the multiplier $\lambda(t)$, say, and the constraint must be expanded into Fourier series, too. Similar considerations hold for distributed constraints. (In connection with perturbation techniques nonlinear boundary conditions can be expanded into series with respect to the small parameters; cf. Mitropolsky, Moseenkov cited in section 15.32).

15.5 Damped nonlinear vibrations

The equations of motion can be reduced to a

svstem of ordinary differential equations as outlined in section 15.12. (Nonlinear boundary condition can frequently trea ed by some special tricks). If the system is dissipative (pas sive), the oscillations will decay, no periodic solution will exist. Small parameter expansions, especially, the asymptotic methods of Krylov, Bogoliubov, Mitropolsky are applicable.

If the system is not dissipative (if it is active), periodic solutions (may) exist. For instance, a rotating shaft with internal friction can oscillate (periodically) beyond the stable parameter regions if nonlinearities are taken into account. Those periodic motions can be calculated numerically as described in section 15.4. Transients might be investigated by asymptotic methods.

15. 6 Stability

The (periodic) solutions of a problem (obtain ed by one of the methods described above) need not be stable. The stability corresponding to Lyapunov's definitions, cf. section 12, has to be investigated separately. In most cases it will be impossible to find a Lyapunov's function. Approximate investigations by means of a (linearized) variational equation of the corresponding ordinary differential equation, cf. (E) in section 15.12., via Floquet solutions, cf. section 14.2, are very cumbersome. Again asymptotic methods are applicable.

16. Nonlinear vibrations of nonautonomous systems

16. 1 Nonlinear transversal vibrations of a rock

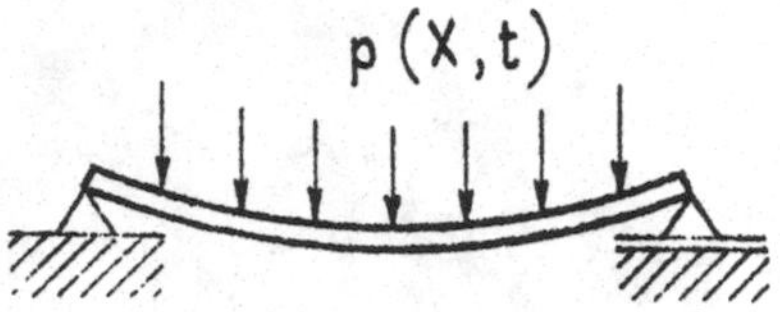

Fig. 16. 1

We assume an uniform Euler-Bernoulli beam (originally plane cross-sectional areas remain plain) and a nonlinear stress-strain relation

$$\sigma = E\varepsilon\left(1 + \alpha\varepsilon^2\right) - \text{conservative}$$

σ - stress

ε - strain

After some calculation we obtain the equation of motion, in nondimensional variables,

$$Du = \ddot{u} + u^{IV} + \beta\left[u^{IV}u'' + 2u'''^{2}\right]u'' - p(X,t) = 0 \qquad (*)$$

(cf. Kauderer's book cited in section 15. 1). The boundary conditions are, cf. Fig. 16. 1,

$$u(0,t) = u''(0,t) = u(1,t) = u''(1,t) = 0 \qquad (**)$$

16.11 Reduction of (*) to a system of ordinary differential equations

Again we can solve the corresponding linearized (homogeneous) problem and obtain the eigenfunctions $U_n(X)$ and the natural frequencies ω_n, cf. section 15.11. Since we have linear boundary conditions an approximate solution of the form

$$u_A = \sum_{n=1}^{N} a_n(t)\, U_n(X) \qquad \text{(A)}$$

satisfies those boundary conditions and by Galerkin's method we find

$$0 = (Du_A, U_n) = \ddot{a}_n + \omega_n^2 a_n + \beta f_n(a_1, \dots, a_N) - q_n(t) = 0\,, \qquad \text{(B)}$$

in vector notation :

$$\ddot{\underset{\sim}{a}} + \underset{\approx}{C}\underset{\sim}{a} + \beta \underset{\sim}{f}(\underset{\sim}{a}) - \underset{\sim}{q}(t) = 0\,. \qquad \text{(C)}$$

Except for the forcing terms $q_n(t)$ these equations coincide with the equations (D) and (E) from section 15.12. In general, it may happen that the (nonlinear) function $\underset{\sim}{f}$ will depend on t explicitly,

$$\ddot{\underset{\sim}{a}} + \underset{\approx}{C}\underset{\sim}{a} + \beta \underset{\sim}{f}(\underset{\sim}{a}, t) - \underset{\sim}{q}(t) = 0\,, \qquad \text{(D)}$$

for instance in the case of the parametrically excited systems discussed in section 14 ($\underset{\sim}{f}$ may include linear terms $\underset{\sim}{a}$).

16. 2 First order approximation N = 1

A first order approximation

$$u = a(t) \sin \pi X$$

for our special example leads to a Duffing equation with a forcing term :

$$\ddot{a} + a + \beta a^3 = q(t) \qquad \text{(E)}$$

(a , t , q are multiplied by appropriate constants to simplify the form of (E)).

No general solutions of (E) are available. Naturally, "solutions" $a(t)$ can be chosen and the corresponding $q(t)$ can be calculated from (E). Sometimes $q(t) = \zeta a(t) + \eta a^3(t)$ is introduced into (E), where ζ , η are constants. This case can be reduced to the autonomous problem, see section 15. 1, and solved by means of elliptic functions. (In fact this problem is autonomous and does not show the phenomena which occur for "realistic" forcing functions $q(t)$)). (E) can be solved numerically for finite time intervals. Approximate procedures, series expansions etc. are possible.

Of special interest are periodic solutions, $a(t)$, for periodic forcing functions, $q(t)$. We discuss the widely treated case $q(t) = q_0 \cos \Omega t$:

$$Ea \equiv \ddot{a} + \delta \dot{a} + a + \beta a^3 - q_0 \cos \Omega t = 0 \qquad \text{(F)}$$

(To show some phenomena we include a linear damping term in (F), δ -coefficient). Differential equations which are periodic with respect to the explicitely contained time are called "periodic differential equations".

16.21 One term Fourier approximation

As a rough approximate solution of (F) we choose

$$a = A \cos(\Omega t - \varphi) .$$

By Fourier expansion we obtain

$$E\left[A \cos(\Omega t - \varphi)\right] = F_{1s}(A, \varphi, \Omega) \sin \Omega t + F_{1c}(A, \varphi, \Omega) \cos \Omega t + \text{h.o.t.}$$

(h. o. t. -higher order terms).

Omitting the h. o. t. we get from $F_{1s} = 0$, $F_{1c} = 0$

$$\tan \varphi = \frac{\Omega \delta}{1 + \frac{3}{4} \beta A^2 - \Omega^2}$$

$$\left[1 + \frac{3}{4} \beta A^2 - \Omega^2\right]^2 + 4 \delta^2 = \frac{q_0^2}{A^2} . \qquad \text{(G)}$$

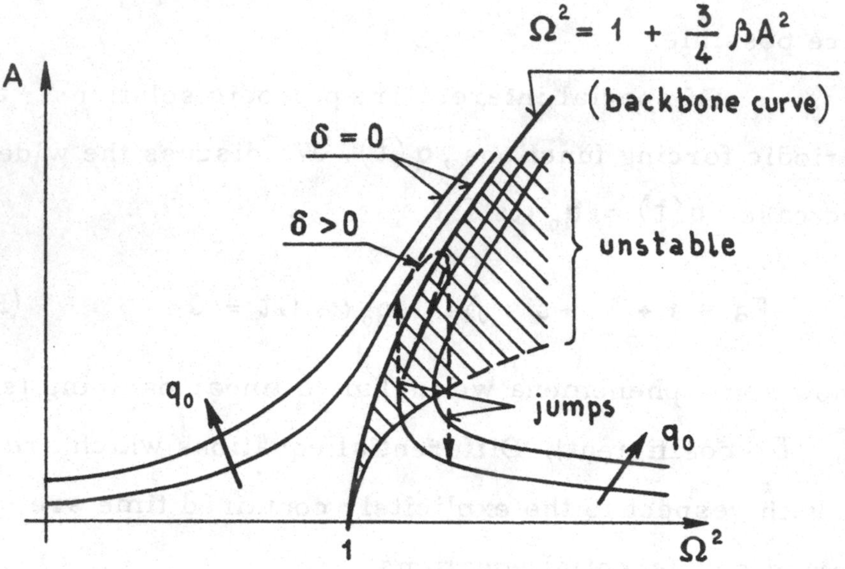

Fig. 16.2

Fig. 16.2 shows the response curves. We see

1. No resonance (infinite A).
2. Two stable solutions in certain Ω regions.
3. Jump phenomena in the vicinity of $\Omega = 1$ when Ω is increased or decreased slowly.

For some more careful investigations on jump phenomena and for response curves of other differential equations see Bogoliubov, Mitropolsky's book cited in section 15.32.

The following investigations will show us some of the shortcomings of the one term Fourier approximation.

16.22 Series expansion; perturbation techniques for periodic equations

For simplicity we assume $\delta = 0$:

$$\ddot{a} + a + \beta a^3 = q_0 \cos \Omega t . \qquad \text{(H)}$$

β is a small parameter; (H) is a quasi linear differential equation.

16.221 Non-resonance case, $|\Omega - 1| \not\ll 1$.

Series expansion with respect to β ,

$$a = a^{(0)} + \beta a^{(1)} + \beta^2 a^{(2)} + \dots , \qquad \text{(I)}$$

leads to

$$\left.\begin{aligned} \beta^0 &: \ddot{a}^{(0)} + a^{(0)} = q_0 \cos \Omega t \\ \beta^1 &: \ddot{a}^{(1)} + a^{(1)} = -\beta \left(a^{(0)}\right)^3 . \\ &\ \vdots \end{aligned}\right\} \qquad \text{(J)}$$

Solutions :

$$a^{(0)} = \frac{q_0}{1-\Omega^2} \cos \Omega t$$

$$a^{(1)} = -\frac{3}{4} \frac{q_0^3}{(1-\Omega^2)^4} \cos \Omega t - \frac{1}{4} \frac{q_0^3}{(1-\Omega^2)^3 (1-9\Omega^2)} \cos 3\Omega t$$

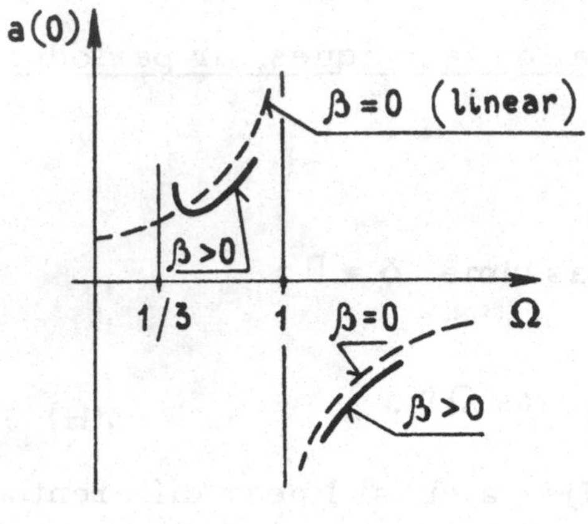

Fig. 16.3

We see: The series expansion shows resonance phenomena at $\Omega = 1$ and at $\Omega = 1/3$. (Higher order approximations show resonance for $\Omega = 1/n$, $n = 1, 3, 5, \dots$).

16.222 Resonance case; $|\Omega - 1| \ll 1$.

An expansion similar to equations (I), (J) in section 16.221 is impossible. At least the term $q_0 \cos \Omega t$ must be shifted into the second equation of (J): To achieve that we assume $\beta > 0$ and introduce

$$a = \bar{a} / \varepsilon$$

into the equation (H), where

$$\varepsilon = \sqrt[3]{\beta}$$

We obtain

$$\ddot{\bar{a}} + \bar{a} + \varepsilon \bar{a}^3 = \varepsilon q_0 \cos \Omega t \qquad \text{(K)}$$

An expansion similar to (I) leads to

$$\varepsilon^0: \quad \ddot{\bar{a}}^{(0)} + \bar{a}^{(0)} = 0$$
$$\varepsilon^1: \quad \ddot{\bar{a}}^{(1)} + \bar{a}^{(1)} = q_0 \cos \Omega t - (\bar{a}^{(0)})^3$$
$$\vdots$$

Assuming $\bar{a}^{(0)} = A \cos t$ as generating solution we cannot prevent resonance in the second equation. To avoid this difficulty we apply Lindstedt's idea. We introduce $\tau = \Omega t$ into (K) and obtain

$$\Omega^2 \ddot{\bar{a}} + \bar{a} + \varepsilon \bar{a}^3 = \varepsilon q_0 \cos \tau \,. \qquad \text{(L)}$$

($\ddot{\bar{a}} = d^2\bar{a} / d\tau^2$). We are looking for 2π-periodic solutions $\bar{a}(\tau + 2\pi) = \bar{a}(\tau)$. A series expansion, cf. (J) in section 15.22,

$$\Omega^2 = 1 + \varkappa_1 \varepsilon + \varkappa_2 \varepsilon^2 + \dots \,,$$
$$\bar{a} = \bar{a}^{(0)} + \varepsilon \bar{a}^{(1)} + \varepsilon^2 \bar{a}^{(2)} + \dots \,,$$

leads to

$$\left.\begin{aligned} \varepsilon^0: \quad & \ddot{\bar{a}}^{(0)} + \bar{a}^{(0)} = 0 \\ \varepsilon^1: \quad & \ddot{\bar{a}}^{(1)} + \bar{a}^{(1)} = q_0 \cos \tau - (\bar{a}^{(0)})^3 - \varkappa_1 \ddot{\bar{a}}^{(0)} \\ & \vdots \end{aligned}\right\} \qquad \text{(M)}$$

As generating solution we choose

$$\bar{a}^{(0)} = A \cos \tau \,.$$

For the higher order terms, $k > 0$, we assume that the $\bar{a}^{(k)}(\tau)$ do not contain terms $\cos \tau$, $\sin \tau$. (The term $A \cos \tau$ in $\bar{a}^{(0)}$ represents the complete first Fourier term of $\bar{a}(\tau)$). A different choice might be

$$\bar{a}^{(k)}(0) = 0 \,, \quad \dot{\bar{a}}^{(k)}(0) = 0 \quad \text{for } k > 0 \,.$$

Note that every equation $k > 0$ of (M) contains a term $\varkappa_k \ddot{\bar{a}}^{(0)}$. Thus, resonance can be prevented by a proper choice of $\varkappa_k$.

We obtain from (M)

$$\bar{a}^{(0)} = A \cos\tau\ , \quad \bar{a}^{(1)} = \frac{A^3}{32}\cos 3\tau\ , \ldots$$

$$\varkappa_1 = \frac{3}{4}A^2 - \frac{q_0}{A}\ , \ldots$$

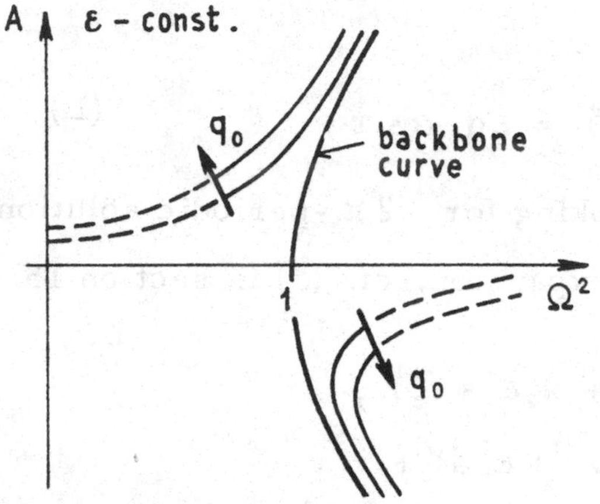

Fig. 16.4 shows $\Omega^2(\varepsilon, q_0, A)$, cf. Fig. 16.2

Fig. 16.4

If we characterize the 2π-periodic (odd harmonic) solutions $\bar{a}(\tau, \varepsilon, \Omega, q_0)$ of (L) by their initial values $\bar{a}(0, \varepsilon, \Omega, q_0)$, $\dot{\bar{a}}(0, \varepsilon, \Omega, q_0) = 0$, we obtain by the procedure outlined above a parameter representation:

$$\bar{a}(0, \varepsilon, A) = A + \varepsilon\frac{A^3}{32} + \ldots\ ,$$

$$\Omega^2(\varepsilon, q_0, A) = 1 + \varepsilon\left[\frac{3}{4}A^2 - \frac{q_0}{A}\right] + \ldots\ ;$$

where A is the parameter.

16.223 Superharmonic resonance; $|n\Omega - 1| \ll 1$, $n = 3,5,\ldots$

Similar expansions as in section 16.222 can be used to investigate the solutions $a(t)$ of (H) in the vicinity of $\Omega = 1/n, n = 3,5,\ldots$. In the neighbourhood of $\Omega = 1/3$, say, the Fourier series for $a(t)$ contains a very large term $A_3 \cos 3\Omega t$ which may dominate all the other terms of the Fourier series. Therefore such solutions are frequently looked upon as "super harmonic" response of the system (the period of the response, $T_{resp} = 2\pi / 3\Omega$, is a fraction of the period of the forcing function, $T_{forc.} = 2\pi/\Omega$).

But a differential equation of the form

$$F(\ddot{a}, \dot{a}, a) = q(t) \tag{N}$$

where F is an analytic function of its arguments and

$$q(t + T_{forc.}) = q(t), \quad T_{forc} \text{ - least period,}$$

can not possess solutions $a(t)$ which have a period, $T_{resp.}$,

$$a(t + T_{resp.}) = a(t), \tag{O}$$

which is less than $T_{resp.}$,

$$0 < T_{resp.} < T_{forc},$$

since substitution of (O) into (N) leads to a contradiction.

Example for a differential equation which has a superharmonic solution :

$$(\ddot{a} + a + \beta a^3 - q_0 \cos \Omega t)(\ddot{a} + 9\Omega^2 a) = 0$$

has the solutions

$$a = A \cos (3\Omega t + \varphi)$$

A , φ arbitrary.

However, (N) may have solutions for which $T_{resp.} = n\, T_{forc.}$, n-integer. We shall construct an example for such a "subharmonic" solution in the next section.

16.224 Subharmonic solutions

We start from the equation (H) and introduce $\Omega t = \tau$:

$$\Omega^2 \ddot{a} + a + \beta a^3 = q_0 \cos \tau \ , \quad \ddot{a} = d^2 a / d\tau^2. \tag{P}$$

Now, we expand Ω^2 and a with respect to β :

$$\left.\begin{aligned} \Omega^2 &= \Omega_0^2 + \varkappa_1 \beta + \varkappa_2 \beta^2 + \ldots , \\ a &= a^{(0)} + \beta a^{(1)} + \beta^2 a^{(2)} + \ldots . \end{aligned}\right\} \tag{Q}$$

Putting (Q) in (P) we obtain

$$\left.\begin{aligned} \beta^0 &: \ \Omega_0^2 \ddot{a}^{(0)} + a^{(0)} = q_0 \cos \tau \\ \beta^1 &: \ \Omega_0^2 \ddot{a}^{(1)} + a^{(1)} = -(a^{(0)})^3 - \varkappa_1 \ddot{a}^{(0)} \\ &\ \ \vdots \end{aligned}\right\} \tag{R}$$

We choose the generating solution

$$a^{(0)} = \frac{q_0}{1-\Omega_0^2} \cos \tau + A \cos \frac{1}{\Omega_0} \tau .$$

For $\Omega_0 = 2$ we get

$$a^{(0)} = -\frac{q_0}{3} \cos \tau + A \cos \frac{\tau}{2} .$$

"No resonance" in the second equation of (R) leads to

$$\varkappa_1 = 3A^2 + \frac{2}{3} q_0^2 .$$

Fig. 16. 5 shows the corresponding response curves. At $a(0) = -q_0/3 + \ldots$ and $\Omega^2 = 4 + 2\beta q_0^2/3 + \ldots$ a 4π-periodic solution of (P) - subharmonic solution of the second order - branches off from the original 2π-periodic solution. Higher order expansions show that "super-subharmonic" solutions branch off from the original 2π-periodic solution in the vicinity of all rational values $\Omega_0 = n/m$, n, m-integers. If damping is included most of these (but not all!) bifurcation points disappear. (That need not mean that the corresponding "branched off" solutions must disappear, the solutions may just separate from each other). It can be proved that (P) possesses an infinite number of periodic solutions, but if damping is taken into account there exists only a finite number of periodic solutions.

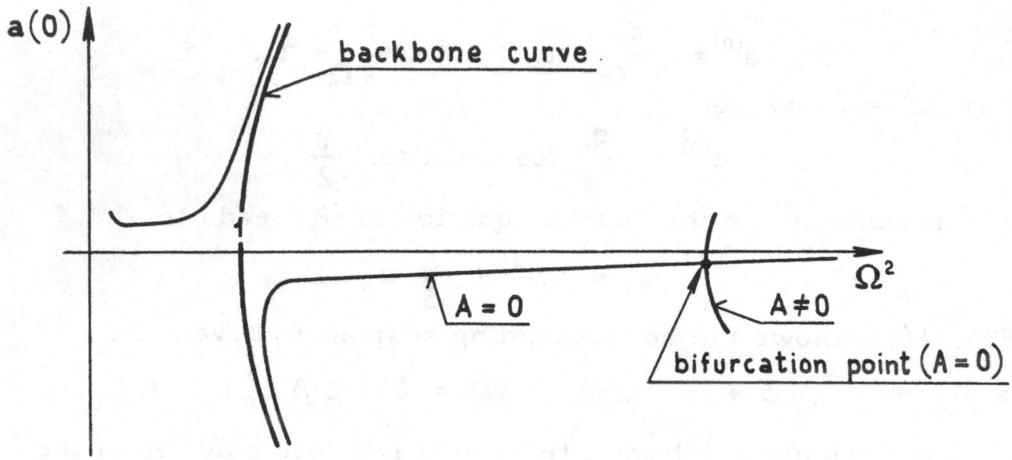

Fig. 16. 5

16. 23 Some periodic solutions obtained numerically for Duffing's equation

Fig. 16. 6 shows some initial values $a(0), (\dot{a}(0) = 0)$, for $2\pi/\Omega$-periodic solutions of

$$\ddot{a} + a + \beta a^3 = q_0 \cos \Omega t$$

for $\beta = 1$ and $q_0 = 0{,}2$. At $\Omega = 1/n$, $n = 3, 5, \dots$, we find the resonances predicted by the series expansion, cf. section 16. 221. In the vicinity of $\Omega = 1/n$, $n = 2, 4, \dots$, the response curves have unstable regions. (High order series expansions would be necessary to show this phenomenon analytically). At the points B and C, see Fig. 16.6,

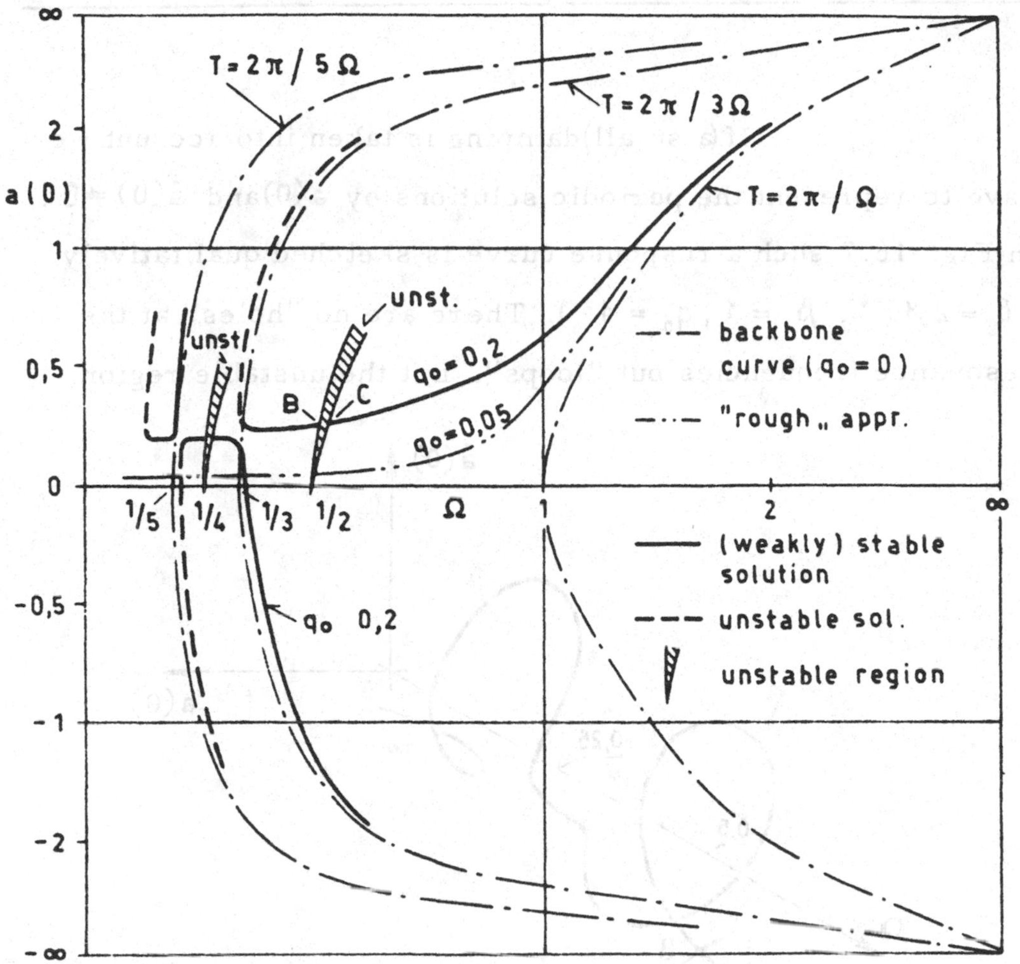

Fig. 16.6

$2\pi/\Omega$ periodic solutions $a_I(t), a_{II}(t)$, $a_{III}(t)$, $a_{IV}(t)$ branch off. These solutions have stable regions. a_I, a_{II}, which branch off from B satisfy $a_{II}(t) = - a_I(t + \pi/\Omega)$, they are even with respect to $t = 0$ and $t = \pi/\Omega$. a_{III}, a_{IV} branch of from C, satisfy $a_{IV}(t) = - a_{III}(t + \pi/\Omega)$, and are odd with respect to $t = \pi/2\Omega$ and $t = 3\pi/2\Omega$. (It is possible that there exist more unstable regions).

If (a small) damping is taken into account we have to represent the periodic solutions by $a(0)$ and $\dot{a}(0) \neq 0$. In Fig. 16.7 such a response curve is sketched qualitatively ($\delta = 2\cdot 10^{-4}$, $\beta = 1$, $q_0 = 1$). There are no "holes" at the resonance frequencies but "loops". But the unstable region

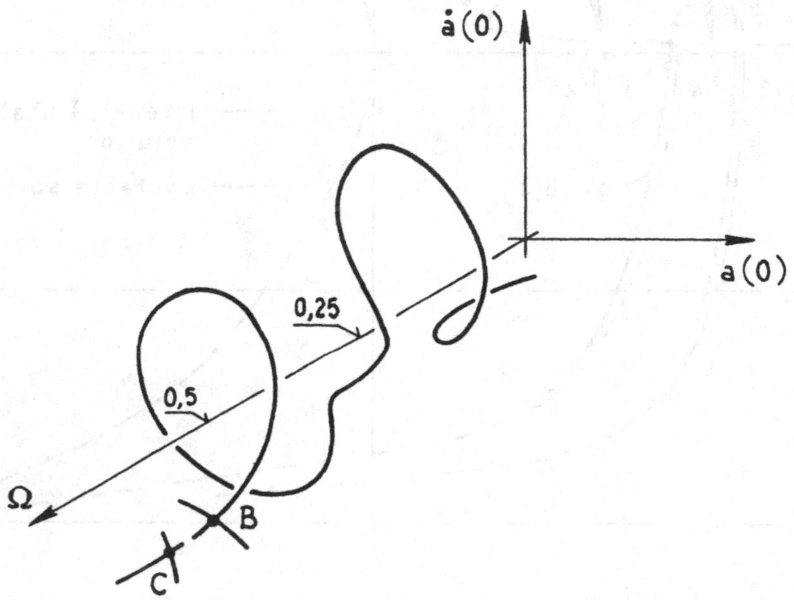

Fig. 16.7

in the vicinity of $\Omega = 1/2$ is preserved. B and C are still bifurcation points. (There may exist some more very narrow unstable regions).

16.24 Almost periodic equations

In section 16.22 we investigated periodic

forced vibrations, cf. equations (E) and (F). Frequently, the forcing functions are not periodic but can be represented by a finite or infinite sum of Fourier terms, e. g.,

$$q(t) = q_1 \cos \omega_1 t + q_2 \cos \omega_2 t \,. \qquad (S)$$

If the ratio

$$\frac{\omega_1}{\omega_2} = \frac{n}{m}$$

is rational (n , m are relatively prime integers), then (S) is periodic; the least period of q(t) is

$$T = \frac{2\pi}{\omega_1} n = \frac{2\pi}{\omega_2} m \,.$$

If ω_1 / ω_2 is irrational,(S) represents an almost periodic function.

16. 241 Almost periodic functions (a. p. f.)

Harald Bohr defined :

Let q(t) be a continuous function defined for all values t from $-\infty < t < \infty$ Then $q(t)$ is called an a. p. f. if for any arbi trarily small positive number ε a positive number $l(\varepsilon)$ can be found such that within each intervall of the length l at least one number $\tau(\varepsilon)$ can be found for which for all t the inequality

$$|q(t+\tau) - q(t)| < \varepsilon$$

is satisfied.

Some basic properties of a. p. f.

1. Every a. p. f. is bounded on $-\infty < t < \infty$.
2. Every a. p. f. is uniformly continuous on $-\infty < t < \infty$.
3. A finite sum of a. p. f. is an a. p. f.
4. The product of two a. p. f. is an a. p. f.
5. If g(t) is an a. p. f., and if $g(t) > 0$ for $-\infty < t < \infty$, then $1/g$ is an a. p. f.
6. The limit $g(t)$ of a uniformly converging sequence of a. p. f. $g_1(t)$, $g_2(t)$, ..., is an a. p. f.

A periodic function is a special case of an a. p. f. Thus, the finite sum

$$q(t) = A_0 + \sum_{j=1}^{J} (A_j \cos \omega_j t + B_j \sin \omega_j t)$$

is an a. p. f. for arbitrary (real) numbers ω_j .

Any a. p. f. can be expanded into a (generalized) Fourier series

$$q(t) = A_0 + \sum_{j=1}^{\infty} (A_j \cos \omega_j t + B_j \sin \omega_j t) . \qquad \text{(T)}$$

(cf. A.S. Besicovich, Almost periodic functions, Cambridge 1938; reprinted by Dover, New York).

Let $Q(t_1, t_2, \dots, t_K)$ be 2π-periodic with respect to each of its arguments,

$$Q(t_1, t_2, \dots, t_K) = Q(t_1+2\pi, t_2, \dots, t_K) = Q(t_1, t_2+2\pi, \dots t_K) = \dots .$$

The special a. p. f.

$$q(t) = Q(\nu_1 t, \nu_2 t, \dots, \nu_K t) \tag{U}$$

is a <u>quasi periodic function.</u> The (real) numbers ν_k, $k = 1, \dots, K$, are called <u>basis frequencies</u>. The ω_j in the Fourier expansion (T) of (U) are linear combinations of the ν_k,

$$\omega_j = m_{j1}\nu_1 + m_{j2}\nu_2 + \dots + m_{jK}\nu_K ,$$

with integer numbers m_{jk} (In general, an a. p. f. has an infinite basis).

16. 242 Break down of the perturbation technique for a. p. differential equations

The differential equation (D), section 16. 11, is called almost periodic if $\underset{\sim}{f}(\underset{\sim}{a}, t)$ and/or $\underset{\sim}{q}(t)$ are almost periodic with respect to t.

We try to find an a. p. solution of the special quasiperiodic differential equation

$$\ddot{a} + a + \beta a^3 = \beta(q_1 \sin t + q_2 \sin \omega t) \tag{V}$$

by a series expansion with respect to β,

(cf. Malkin's book cited in section 15.22, and N. Minorsky, Nonlinear oscillations, Van Nostrand, New York 1962).

Introducing

$$a = a^{(0)} + \beta a^{(1)} + \beta^2 a^{(2)} + \dots$$

into (V) we obtain

$$\ddot{a}^{(0)} + a^{(0)} = 0$$

$$\ddot{a}^{(1)} + a^{(1)} = q_1 \sin t + q_2 \sin \omega t - (a^{(0)})^3$$

$$\ddot{a}^{(2)} + a^{(2)} = -3 a^{(1)} (a^{(0)})^2$$

$$\vdots$$

As generating solution we choose

$$a^{(0)} = B_0 \sin t$$

To avoid resonance in $a^{(1)}$ we require

$$B_0^3 = 4 q_1 / 3 \, .$$

We obtain

$$a^{(1)} = -\frac{B_0^3}{32} \sin 3t + \frac{q_2}{1-\omega^2} \sin \omega t + B_1 \sin t .$$

To avoid resonance in $a^{(2)}$ $B_1 = -B_0^3 / 48$ is chosen, we obtain

$$a^{(2)} = -\frac{3 B_0^2 q_1}{1024} \sin 3t + \frac{3 B_0^2 q_1}{2048} \sin 5t - \frac{3 B_0^2 q_2}{2(1-\omega^2)^2} \sin \omega t$$

$$+ \frac{3 B_0^2 q_2 \sin(\omega - 2) t}{2(1-\omega^2)\left[1-(\omega-2)^2\right]} + \frac{3 B_0^2 q_2 \sin(\omega + 2) t}{2(1-\omega^2)\left[1-(\omega+2)^2\right]} + B_2 \sin t$$

etc.

This is a formal series expansion. Higher $a^{(k)}$ will contain terms

$$\cdots \frac{\sin(m\omega - nt)}{\left[1-(m\omega - n)^2\right]}$$

Since ω is irrational it may happen that $\left|1-(m\omega-n)^2\right| \ll 1$ for large integers n, m. Thus, it may happen that the higher terms contain "small divisors", the corresponding $a^{(k)}$ become very large, the series for $a(t)$ does not converge.

Special expansion techniques, due to Krylov and Bogoliubov, are described in the books of Malkin and Mitropolsky.

16.243 Solution of a. p. differential equations by the method of successive approximations

16.2431 Linear a. p. differential equations

The linear a. p. differential equation

$$\ddot{a} + \delta\dot{a} + a = q(t), \qquad 0 < \delta < 1/2,$$

where $q(t)$ is an a. p. f., has the solution

$$a(t) = \frac{1}{\nu}\int_{-\infty}^{t} e^{-\delta(t-\sigma)/2} \sin\nu(t-\sigma)\, q(\sigma)\, d\sigma, \qquad \text{(W)}$$

$$\nu = \sqrt{1-\delta^2/4}$$

The right hand side of (W) is an a. p. f. :

$$|a(t+\tau)-a(t)| =$$

$$\left|\frac{1}{\nu}\int_{-\infty}^{t+\tau} e^{-\delta(t+\tau-\sigma)/2}\sin\nu(t+\tau-\sigma)\,q(\sigma)d\sigma - \frac{1}{\nu}\int_{-\infty}^{t} e^{-\delta(t-\sigma)/2}\sin\nu(t-\sigma)q(\sigma)d\sigma\right|$$

$$= \frac{1}{\nu}\left|\int_{-\infty}^{t}[q(\sigma+\tau)-q(\sigma)]\; e^{-\delta(t-\sigma)/2}\sin\nu(t-\sigma)\,d\sigma\right| < \varepsilon$$

if $$|q(\sigma+\tau)-q(\sigma)| < 2\varepsilon\nu/\delta\,.$$

Furthermore,

$$|a(t)| < 2M/\delta\nu\,,$$

where

$$M = \max_t |q(t)|$$

16.2432 Nonlinear a. p. differential equations

Quasilinear differential equations :

$$\ddot{a} + \delta\dot{a} + a + \beta f(a,t) = q(t)\,,$$

where $0 < \delta < 1/2$, $q(t)$ is an a. p. f. The function $f(a,t)$ i almost periodic with respect to t (uniformly with respect to a), furthermore $f(a,t)$ satisfies a Lipschitz condition with respect to a ,

$$|f(a_1,t) - f(a_2,t)| < L|a_1 - a_2|\,,$$

L does not depend on t

Successive approximations :

$$a^{(0)} = \frac{1}{\nu}\int_{-\infty}^{t} e^{-\delta(t-\sigma)/2} \sin\nu(t-\sigma)\, q(\sigma)\, d\sigma ,$$

$$a^{(k+1)} = \frac{1}{\nu}\int_{-\infty}^{t} e^{-\delta(t-\sigma)/2} \sin\nu(t-\sigma) \left[q(\sigma) - \beta f(a^{(k)}(\sigma), \sigma)\right] d\sigma ,$$

$$k = 0, 1, \ldots .$$

All $a^{(k)}(t)$ are a. p. f. The $a^{(k)}$ converge to an a. p. solution, $a(t)$, if β is small enough.

16. 3 Higher order approximations $N > 1$

The methods outlined in 16. 2 for quasilinear periodic and almost periodic differential equations are applicable to systems of such equations, too. In many problems difficulties arise from internal resonances and small divisors.

16. 4 Numerical computation of special solutions

Periodic solutions of periodic equations of motion can be calculated by Urabe's version of Galerkin's procedure as discussed in section 15. 4.

16.5 Stability

The stability of the periodic, almost periodic, etc. solutions has to be investigated. For periodic solutions the procedure mentioned in section 15.6 is applicable. The stability of almost periodic solutions of quasilinear differential equations can be investigated by analytic methods (see Malkin's book cited above).

17. Asymptotic methods

In this section we shall discuss some methods, originally used in celestial mechanics, which were applied to special technical problems first by van der Pol. Krylov and Bogoliubov, and many others, extended and generalized these methods. (See the survey article by Y. A. Mitropolsky, Averaging method in nonlinear mechanics, Internat. J. of Nonlin. Mech. 2 (1967) 69 - 95).

17. 1 The method of the slowly varying amplitudes

We resume the example of the hinged-hinged bar which is loaded axially by a pulsating force, cf. section 14. 1, but this time we take some nonlinearities into account. (This problem was solved by F. Weidenhammer, Das Stabilitatsverhalten der nichtlinearen Biegeschwingungen des axial pulsiernd belasteten Stabes, Ing.-Arch. 24 (1956) 53-68).

17. 11 Equation of Motion

Uniform Euler-Bernoulli beam; μ, EI, A-constant (A- cross-sectional area). The longitudinal extension is taken into account.

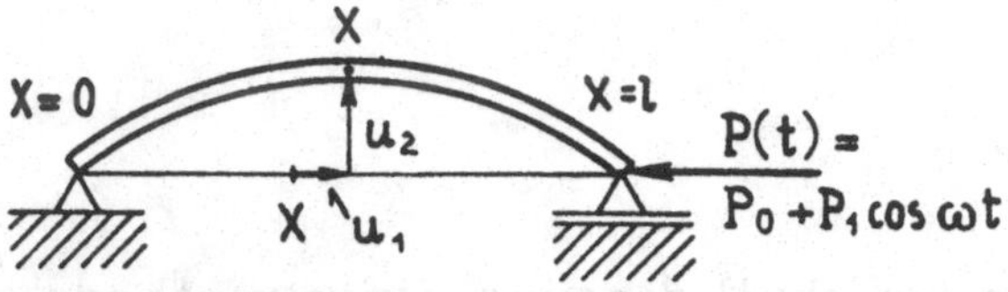

Fig. 17.1

The strain of the center line of the beam is approximated by (cf. Kauderer's book cited in sect. 15.1):

$$\varepsilon_0 = \sqrt{1 + 2u_1' + u_1'^2 + u_2'^2} - 1 \approx u_1' + \frac{1}{2} u_2'^2$$

The longitudinal strain of an element which has the distance Y from the centerline is assumed to be

$$\varepsilon_X = \varepsilon_0 - Y u_2'' .$$

Introducing Hooke's law, neglecting the inertia terms $\ddot{u}_1$, approximating the longitudinal motion $u_1(X, t)$ bei $X\, U_1(t)$, which is allowed for excitations far below the first longitudinal natural freqency, and taking one term $u_2(X,t) = a(t) \sin \pi X / l$ into account, Weidenhammer obtains a single nonlinear second order ordinary differential equation for $a(t)$:

$$\omega^2 \ddot{a} + a = -\delta \omega a + p_1 \cos t\, a - a^3 \qquad \text{(A)}$$

(A) is written in a nondimensional form. The first natural frequency is normalized to 1, ω is the (nondimensional) frequency of the pulsating force, $\delta \dot{a}$ represents a linear external damping, p_1 is proportional to P_1 (see Fig. 17.1).

Assuming δ and p_1 to be proportional to a small parameter ε, $\varepsilon\delta \Longrightarrow \delta$, $\varepsilon p_1 \Longrightarrow p_1$, and introducing $a\sqrt{\varepsilon}$ for $a(t)$ we find from (A)

$$\omega^2 \ddot{a} + a = \varepsilon(-\delta\omega\dot{a} + p_1 \cos t \; a - a^3) \tag{B}$$

17. 12 Van der Pol's method

(B) has the trivial solution $a(t) \equiv 0$. In section 14. 12 we saw that $a(t) \equiv 0$ becomes unstable in the vicinity of $\omega = 2$ $(= 2\,\omega_1$ since $\omega_1 = 1)$.
In that neighborhood we expect to find non-trivial periodic solutions of (B) which have the period 4π. Correspondingly we write

$$a(t) = A(t) \cos t/2 + B(t) \sin t/2 \tag{C}$$

or

$$a(t) = Q(t) \cos(t/2 + \varphi(t)) \tag{D}$$

(C) and (D) are equivalent:

$$A = Q \cos\varphi \quad , \quad B = -Q \sin\varphi \,. \tag{E}$$

$A(t)$ and $B(t)$ are introduced as time dependent functions to take into account the deviation of $a(t)$ from a sinusoidal motion ($Q(t)$ and $\varphi(t)$ might serve for the same purpose). Hopefully, the terms $\cos t/2$ and $\sin t/2$ will catch the "fast" parts of the motion and the amplitudes $A(t), B(t)$ will vary slowly (in (D) the amplitude $Q(t)$ and the phase $\varphi(t)$ would vary slowly).

Since we introduce by (C) two arbitrary functions into (B) we are allowed to choose one arbitrary additional condition. We assume

$$\dot{A} \cos t/2 + \dot{B} \cos t/2 = 0 \qquad \text{(F)}$$

Because of (F) we obtain from (C)

$$\left.\begin{aligned} \dot{a} &= - A/2 \ \sin t/2 + B/2 \ \cos t/2 , \\ \ddot{a} &= - \dot{A}/2 \ \sin t/2 + \dot{B}/2 \ \cos t/2 \\ &\quad - A/4 \ \cos t/2 - B/4 \ \sin t/2 . \end{aligned}\right\} \qquad \text{(G)}$$

Substituting (G) into (B) and solving the resulting equation, together with (F), for $\dot{A}$ and $\dot{B}$ we obtain two first order differential equations for $A(t)$ and $B(t)$:

$$\left.\begin{aligned} \frac{\omega^2}{2}\dot{A} &= - A\left(\frac{\omega^2}{4} - 1\right) \cos\frac{t}{2} \sin\frac{t}{2} - B\left(\frac{\omega^2}{4} - 1\right) \sin^2\frac{t}{2} \\ &\quad - \varepsilon F(A,B,t) \sin\frac{t}{2} = R_A(A,B,t) , \\ \frac{\omega^2}{2}\dot{B} &= A\left(\frac{\omega^2}{4} - 1\right) \cos^2\frac{t}{2} + B\left(\frac{\omega^2}{4} - 1\right) \sin\frac{t}{2} \cos\frac{t}{2} \\ &\quad + \varepsilon F(A,B,t) = R_B(A,B,t) , \end{aligned}\right\} \qquad \text{(H)}$$

where

$$F(A,B,t) = -\omega\delta\left(-\frac{A}{2}\sin\frac{t}{2}+\frac{B}{2}\cos\frac{t}{2}\right)+p_1\cos t\left(A\cos\frac{t}{2}+B\sin\frac{t}{2}\right)$$
$$-\left(A\cos\frac{t}{2}+B\sin\frac{t}{2}\right)^3.$$

This set of equations is still "exactly" equivalent to the equation (B).

Since ω is assumed to be close to 2, $|\omega^2/4-1|$ will be of the order ε,

$$|\omega^2/4-1| = O(\varepsilon)$$

Thus, the right hand side of (H) is of the order ε, $A(t)$ and $B(t)$ vary slowly.

Van der Pol's argument: The rapidly changing terms on the right hand side of (H), which depend explicitely on the time, do not contribute much to A and B. By taking the "average" of those terms over their period 4π,

$$\left.\begin{aligned} \omega^2\dot{A}/2 &= \frac{1}{4\pi}\int_0^{4\pi} R_A(A,B,t)\,dt \\ \omega^2\dot{B}/2 &= \frac{1}{4\pi}\int_0^{4\pi} R_B(A,B,t)\,dt \end{aligned}\right\} A,B\text{-"constant"}, \qquad \text{(I)}$$

we obtain the set of two first order nonlinear autonomous differential equations

$$\left.\begin{aligned} \omega^2\dot{A} &= B\left[1-\omega^2/4+\varepsilon p_1/2+3\varepsilon(A^2+B^2)/4\right]-\varepsilon\delta\omega A/2\,, \\ \omega^2\dot{B} &= -\varepsilon\delta\omega B/2-A\left[1-\omega^2/4-\varepsilon p_1/2+3\varepsilon(A^2+B^2)/4\right]. \end{aligned}\right\} \quad \text{(J)}$$

From (J) the time can be eliminated:

$$\frac{dA}{dB}=\frac{B\left[1-\omega^2/4+\varepsilon p_1/2+3\varepsilon(A^2+B^2)/4\right]-\varepsilon\delta\omega A/2}{-\varepsilon\delta\omega B/2-A\left[1-\omega^2/4-\varepsilon p_1/2+3\varepsilon(A^2+B^2)/4\right]}\,. \quad \text{(K)}$$

(K) is a first order differential equation for A(B). It describes a directional field in an A-, B- plane ("phase plane"; cf. e. g., Bogoliubov, Mitropolsky's book cited in section 15. 32).

17. 13 Stationary solutions

Stationary solutions, $A = A_0$-constant, $B = B_0$-constant, are obtained from (J) by assuming $\dot{A} = 0$, $\dot{B} = 0$:

Vanishing damping: We get three different sets of solutions:

1: $A_0 = 0$, $B_0 = 0$, $a(t) \equiv 0$

2: $B_0 = 0$, $A_0^2 = -4\left[1-\omega^2/4-\varepsilon p_1/2\right]/3\varepsilon$, $a(t) = \pm A_0 \cos\frac{t}{2}$

3: $A_0 = 0$, $B_0^2 = -4\left[1-\omega^2/4-\varepsilon p_1/2\right]/3\varepsilon$, $a(t) = \pm B_0 \sin\frac{t}{2}$

Non-vanishing damping:

$A_0 = 0$, $B_0 = 0$ is again a solution. For A_0, $B_0 \neq 0$ we obtain

$$A_0^2 + B_0^2 = \frac{4}{3\varepsilon}\left[\omega^2/4 - 1 \pm \sqrt{\varepsilon^2 p_1^2/4 - \varepsilon^2\delta^2\omega^2/4}\right]$$

In Fig. 17.2 some results are represented

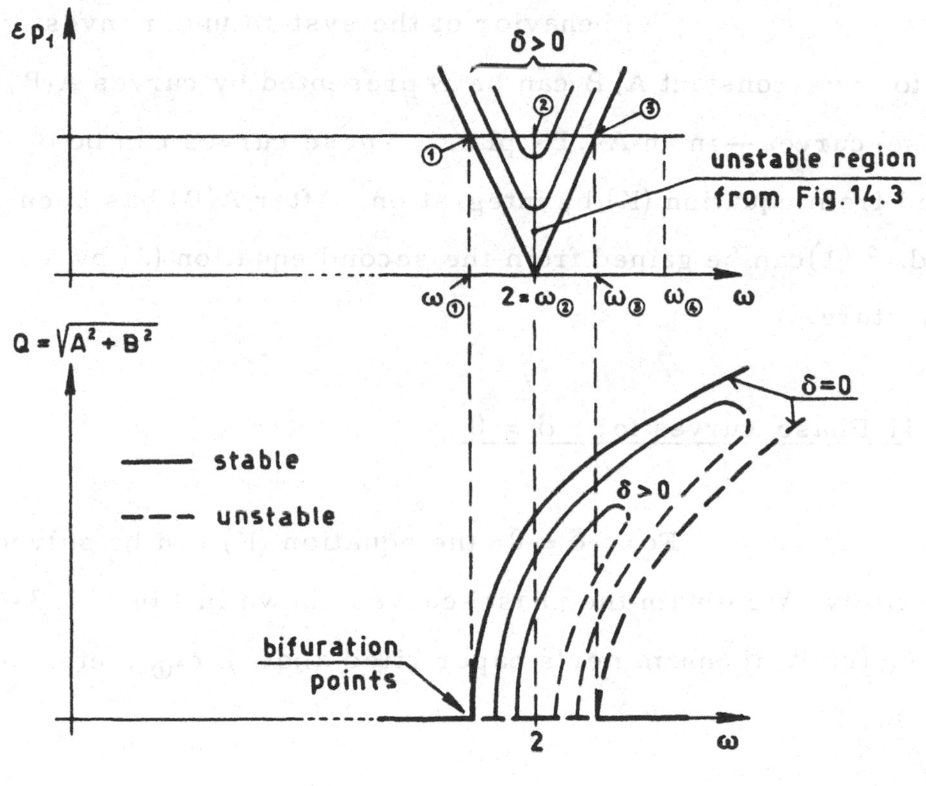

Fig. 17.2

for a fixed value εp_1. In the upper half of that figure the unstable region from Fig.14.3 is sketched, below some response curves are drawn. (The stationary solutions A_0, B_0 are the singular points of the equation (K). The stability of these solutions depends on the character of the singular points.)

17.14 Curves A(B) in an A-, B- plane

The behavior of the system under investigation for non-constant A, B can be represented by curves A(B) —phase-curves— in an A-, B- plane. These curves can be obtained from equation (K) by integration. After A(B) has been found, $B(t)$ can be gained from the second equation (J) by a quadrature.

17.141 Phase-curves for $\delta = 0$

For $\delta = 0$ the equation (K) can be solved explicitely. We obtain the phase-curves shown in Fig. 17.3-17.6 (cf. Weidenhammer's paper cited above). $\omega_{①}$, etc. see Fig. 17.2.

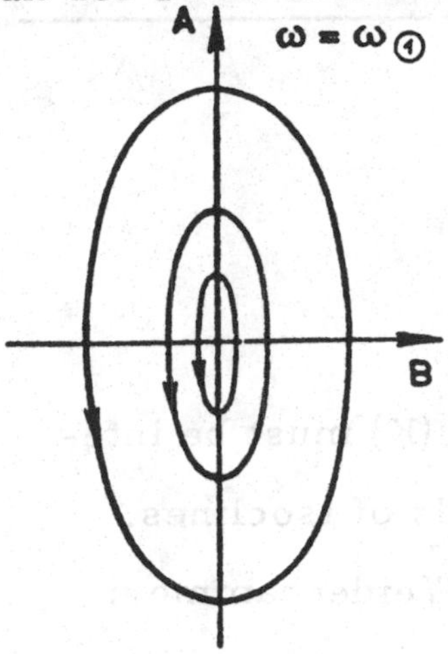

Fig. 17.3

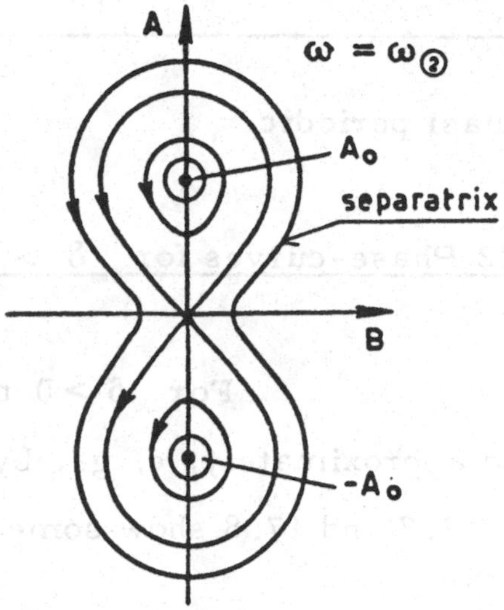

Fig. 17.4

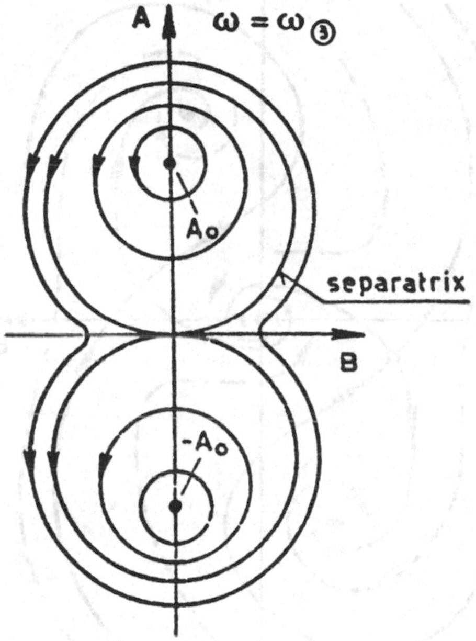

Fig. 17.5

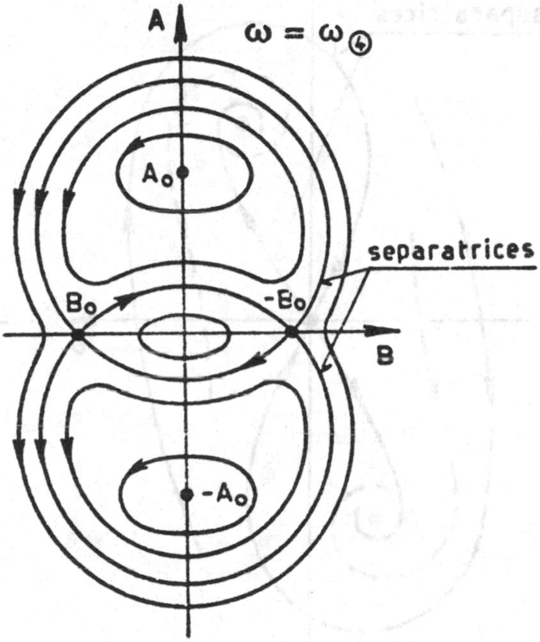

Fig. 17.6

The closed phase-curves (that are not separatrices) represent periodic solutions $A(t), B(t)$ with a period T different from 4π. Thus, in general, the approximate solutions of the form (C) are quasi periodic.

17.142 Phase-curves for $\delta > 0$

For $\delta > 0$ the equation (K) must be integrated approximately, e. g., by the methods of isoclines. Figs. 17.7 and 17.8 show some results of Weidenhammer:

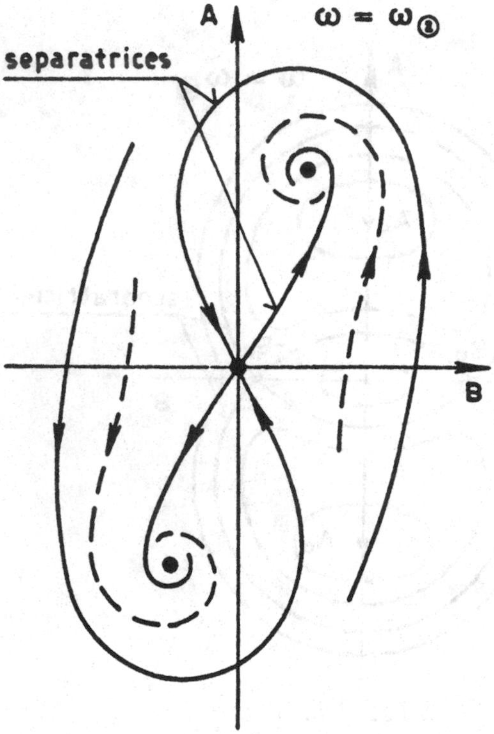

Fig. 17.7

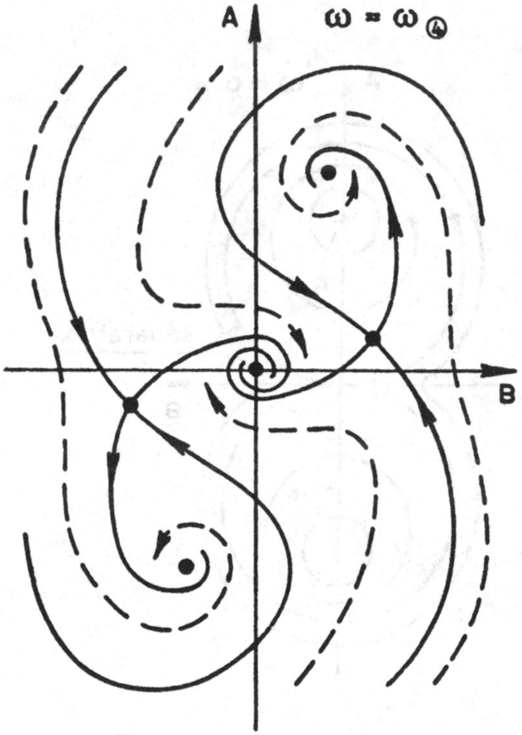

—— separatrices
- - - transients

Fig. 17.8

By these figures the transients of our system can be studied.

17. 2 Generalizations of van der Pol's method

Van der Pol's method was extended into two directions.

First, higher order approximations for (J) were developed which permit to formulate autonomous equations of the form (J) up to an accuracy of the order ε^m: The right hand sides of the nonautonomous equations of the form (H) are expanded by special procedures with respect to ε in such a way that

$$\left.\begin{aligned} \dot{A} &= \varepsilon F_1(A,B) + \varepsilon^2 F_2(A,B) + \cdots + \varepsilon^m F_m(A,B) + \varepsilon^{m+1} r_A(A,B,t), \\ \dot{B} &= \varepsilon G_1(A,B) + \varepsilon^2 G_2(A,B) + \cdots + \varepsilon^m G_m(A,B) + \varepsilon^{m+1} r_B(A,B,t). \end{aligned}\right\} \quad \text{(L)}$$

Then r_A and r_B are neglected; the truncated equations are autonomous. Mostly, (L) is formulated for slowly varying phases and amplitudes, cf. equation (D).

By "up to an accuracy of the order m" is meant that the truncated equations satisfy a relation of the following form

$$\lim_{\varepsilon \to 0} \frac{1}{\varepsilon^m} \left[\dot{A} - \varepsilon F_1 - \cdots - \varepsilon^m F_m \right] = 0 \quad , \ m\text{-fixed}. \qquad \text{(M)}$$

Expansions which satisfy a relation of the form (M) are called asymptotic expansions. For details see Bogoliubov, Mitropolsky's book cited above.

Since the solutions of two first order autonomous equations can always be interpreted graphically, as shown above, some procedures were developed to reduce the investigation of special (restricted) motions of multiple degree of freedom systems to the study of the solutions of such differential equations.

A second extension of Van der Pol's method went into the direction of multiple degree of freedom systems, i. e., to a general method of averaging.

17. 3 The method of averaging

(Cf. Bogoliubov, Mitropolsky's book cited in section 15. 32).

17. 31 Standard form

A quasilinear systems of equations which corresponds to the equation (D) of section16. 11 has the form

$$\ddot{\mathbf{a}} + \mathbf{C}\mathbf{a} = \varepsilon \mathbf{g}(\mathbf{a}, \dot{\mathbf{a}}, t), \tag{N}$$

where $\mathbf{C}$ is assumed to be a positive diagonal matrix and ε is a small parameter. (N) can be replaced by the first order system

$$\dot{\mathbf{y}} + \mathbf{D}\mathbf{y} = \varepsilon \mathbf{h}(\mathbf{y}, t), \tag{O}$$

where

$$\mathbf{y} = \begin{pmatrix} \mathbf{a} \\ \dot{\mathbf{a}} \end{pmatrix}, \quad \mathbf{h} = \begin{pmatrix} 0 \\ \mathbf{g} \end{pmatrix}, \quad \mathbf{D} = \begin{pmatrix} 0 & -\mathbf{I} \\ \mathbf{D} & \mathbf{I} \end{pmatrix}$$

For $\varepsilon = 0$ the equation (O) has the solution

$$\mathbf{y} = e^{t\mathbf{D}} \mathbf{X},$$

where $\mathbf{X} = \mathbf{y}(0)$. (Since $\mathbf{C}$ is a positive diagonal matrix, $\exp(t\mathbf{D})$ is a quasiperiodic matrix.) Variation of $\mathbf{X}$ leads to

$$\dot{\mathbf{X}} = \varepsilon e^{-t\mathbf{D}} \mathbf{h}(e^{t\mathbf{D}} \mathbf{X}, t) = \varepsilon \mathbf{f}(\mathbf{X}, t).$$

$$\dot{\mathbf{X}} = \varepsilon \mathbf{f}(\mathbf{X}, t). \tag{P}$$

The form (P) of the equation of motion is called "standard form $\mathbf{f}(\mathbf{X}, t)$ is assumed to be quasiperiodic or, at least, almost periodic.

17.32 Some formal transformations

Let $\underset{\sim}{F}(\underset{\sim}{X}, t)$ be almost periodic with respect to t and permit a representation of the form

$$\underset{\sim}{F}(\underset{\sim}{X}, t) = \sum_{\nu} e^{i\nu t} \underset{\sim}{F}_{\nu}(\underset{\sim}{X}) .$$

(The ν's are real numbers)

We define

$$\underset{\sim}{F}_0(\underset{\sim}{X}) = \underset{t}{M}\left\{\underset{\sim}{F}(\underset{\sim}{X}, t)\right\} = \lim_{T \to \infty} \frac{1}{T} \int_0^T \underset{\sim}{F}(\underset{\sim}{X}, t)\, dt , \quad \underset{\sim}{X} \text{ fixed,}$$

and

$$\tilde{\underset{\sim}{F}}(\underset{\sim}{X}, t) = \sum_{\nu \neq 0} \frac{e^{i\nu t}}{i\nu} \underset{\sim}{F}_{\nu}(\underset{\sim}{X})$$

We find

$$\frac{\partial \tilde{\underset{\sim}{F}}}{\partial t} = \underset{\sim}{F} - M_t\left\{\underset{\sim}{F}\right\} .$$

17.33 Averaged equation; first approximation

Applying the transformations 17.32 to the equation (P) we obtain

$$\dot{\underset{\sim}{X}} = \varepsilon f_0(\underset{\sim}{X}) + \text{ small sinusoidal terms.}$$

We neglect the small oscillating terms and get the autonomous equation

$$\dot{\underset{\sim}{\xi}} = \varepsilon \underset{\sim}{f}_0 (\underset{\sim}{\xi}) . \tag{Q}$$

The solution $\underset{\sim}{\xi}(t)$ is an approximation to the solution $\underset{\sim}{x}(t)$ of equation (P).

Let $\underset{\sim}{x}$ and $\underset{\sim}{\xi}$ satisfy the same initial conditions, then it can be proved that the error $|\underset{\sim}{\xi} - \underset{\sim}{x}|$ can be made arbitrarily small on an arbitrarily large, but finite, time interval if ε is chosen small enough and if certain restrictions are satisfied.

In general, the solutions of the equation (Q) are more easily obtained and investigated than the solutions of the original equation (P). Frequently, (Q) has a stationary solution $\underset{\sim}{\xi} = \underset{\sim}{\xi}_0$. In that case the (linear) variational equation which may serve to investigate the stability of $\underset{\sim}{\xi}_0$, cf. section 12. 5, is a differential equation with constant coefficients,

$$\Delta \dot{\underset{\sim}{\xi}} = \underset{\sim}{f}_{0\underset{\sim}{\xi}} (\underset{\sim}{\xi}_0) \Delta \underset{\sim}{\xi} . \tag{R}$$

($\underset{\sim}{f}_{0\underset{\sim}{\xi}}$ is the matrix of the partial derivatives $(\partial f_{0i} / \partial \xi_k .)$
The stability of the trivial solution of (R) can be checked by the Hurwitz criteria.

17.34 Order of the first approximation

To show that (Q) is an approximation of the first order, cf. section 17.2, let us define $\underset{\sim}{\xi}(t)$ implicitly by

$$\underset{\sim}{X}(t) = \underset{\sim}{\xi}(t) + \varepsilon \underset{\sim}{\tilde{f}}(t, \underset{\sim}{\xi}) \quad , \text{ cf. section 17.32.} \qquad \text{(S)}$$

Differentiating (S) with respect to t, taking the last relation of section 17.32 into account, we obtain

$$\dot{\underset{\sim}{X}} = \dot{\underset{\sim}{\xi}} + \varepsilon \frac{\partial \underset{\sim}{\tilde{f}}}{\partial \underset{\sim}{\xi}} \dot{\underset{\sim}{\xi}} + \varepsilon \frac{\partial \underset{\sim}{\tilde{f}}}{\partial t} = \left\{ 1 + \varepsilon \frac{\partial \underset{\sim}{\tilde{f}}}{\partial \underset{\sim}{\xi}} \right\} \dot{\underset{\sim}{\xi}} + \varepsilon \underset{\sim}{f}(\underset{\sim}{\xi}, t) + \varepsilon \underset{\sim}{f}_0(\underset{\sim}{\xi}).$$

Putting this into equation (P) and applying the expansions

$$\left\{ 1 + \varepsilon \frac{\partial \underset{\sim}{\tilde{f}}}{\partial \underset{\sim}{\xi}} \right\}^{-1} = 1 - \varepsilon \frac{\partial \underset{\sim}{\tilde{f}}}{\partial \underset{\sim}{\xi}} + \varepsilon^2 \cdots ,$$

and

$$\underset{\sim}{f}(\underset{\sim}{\xi} + \varepsilon \underset{\sim}{\tilde{f}}, t) = \underset{\sim}{f}(\underset{\sim}{\xi}, t) + \varepsilon \frac{\partial \underset{\sim}{f}}{\partial \underset{\sim}{\xi}} \underset{\sim}{\tilde{f}} + \cdots$$

we get

$$\dot{\underset{\sim}{\xi}} = \varepsilon \underset{\sim}{f}_0(\underset{\sim}{\xi}) + \varepsilon^2 \cdots .$$

By similar transformations higher order approximations can

be obtained. (Attention: the relation (S) does not mean that
$|\underset{\sim}{x} - \underline{\xi}| = O(\varepsilon)$ is valid on an <u>infinite</u> time interval!)

18. Investigation of bifurcation points

In sections 16 and 17 we encountered bifurcation points. In diagrams which represent the solutions depending on some parameter e , say, there exist special values of this parameter, e_s , where solutions coincide which for $e \neq e_s$ are of different character. Looked at $e = e_s$ from another point of view we might say that at $e = e_s$ a new solution branches off from the old, known, one; see Fig. 18.1.

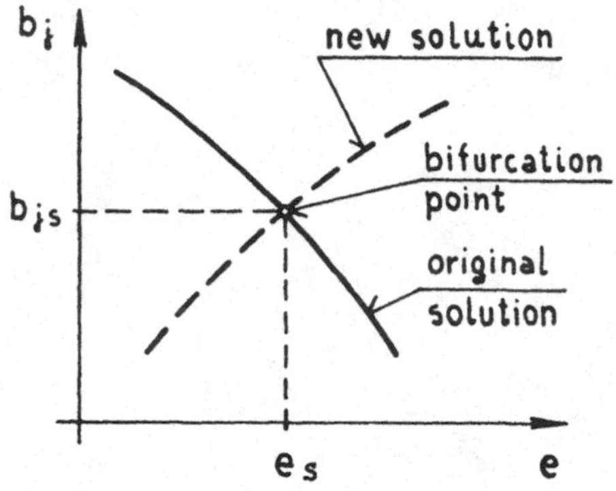

Fig. 18.1

Thus, the investigation of bifurcation points of a known solution may serve to detect new, unknown solutions (new phenomena). We shall outline here a procedure which is applicable to periodic solutions and is connected to the numerical procedure described in section 15.4. In general, these calculations have to be done partly analytical, partly numerical. Frequently, the effort needed to do such investigations is tremendous because numerical difficulties arise.

There exists much literature on the investigation of bifurcation points. We follow mainly E. Schmidt, Ueber die Auflösung der nichtlinearen Integralgleichungen und die Verzweigungen ihrer Lösungen, Math. Annal. 65 (1908) 370-399, and R. G. Bartle, Singular points of functional equations, Trans. Amer. Math. Soc. 75 (1953) 366-384.

18. 1 Location of bifurcation points

In section 15. 4 we established the equation

$$\underset{\sim}{r} = \underset{\sim}{r}(\underset{\sim}{b}) = \mathcal{A} D \mathcal{S} \underset{\sim}{b} = 0 \qquad \text{(A)}$$

from which solutions $\underset{\sim}{b} = \underset{\sim}{b}^{**}$ were obtained by Newton's method. ($\underset{\sim}{r}$ and $\underset{\sim}{b}$ are $MN = L$-vectors, cf. section 15. 4) Let us introduce the parameter e into (A),

$$\underset{\sim}{r}(\underset{\sim}{b}, e) = 0, \qquad \text{(B)}$$

and let $\underset{\sim}{b} = \underset{\sim}{b}(e)$ be a solution of (B)—the asterisks are dropped. By the implicit functions theorem (B) has unique solutions $\underset{\sim}{b}(e)$ in the neighborhood of a fixed value e if the Jacobian matrix of (B) at that point is nonsingular,

$$\det \underset{\sim}{F} \neq 0, \qquad \text{(C)}$$

where

$$\underset{\sim}{F}(e) = \frac{\partial \underset{\sim}{r}}{\partial \underset{\sim}{b}} \ , \quad F_{ik} = \frac{\partial r_i}{\partial b_k} \ , \text{cf. } 15.42 \tag{D}$$

Thus, we can expect to find bifurcation points only at those points $e = e_s$ where $\underset{\sim}{F}$ is singular,

$$\det \underset{\sim}{F}(e_s) = 0 \ ,$$

the corresponding solution $\underset{\sim}{b}(e_s)$ we denote by $\underset{\sim}{b}_s$.
To find all solutions which coincide at $e = e_s$ we have to investigate the manifold of solutions $\underset{\sim}{b}(e)$ in the vicinity of $e = e_s$.

18. 2 Some transformations

Let ε be a small parameter

$$e = e_s + \varepsilon \tag{E}$$

and

$$\underset{\sim}{b}(e_s + \varepsilon) = \underset{\sim}{b}_s + \underset{\sim}{\beta} \tag{F}$$

we have to find $\underset{\sim}{\beta}(\varepsilon)$. Putting (E) and (F) into (B) we obtain

$$\underset{\sim}{r}(\underset{\sim}{b}_s + \underset{\sim}{\beta}, e_s + \varepsilon) = 0 \ ,$$

which we abbreviate by

$$\underset{\sim}{r}_s(\underset{\sim}{\beta}, \varepsilon) = 0 , \tag{G}$$

$$\underset{\sim}{r}_s(0, 0) = 0 . \tag{H}$$

Let us assume that $\underset{\sim}{F}(e_s) = \underset{\sim}{F}_s$ has the rank $L - 1$. Let $\underset{\sim}{p}$ and $\underset{\sim}{q}$ denote the right hand and the left hand eigenvector of $\underset{\sim}{F}_s$, respectively,

$$\left.\begin{aligned} \underset{\sim}{F}_s \underset{\sim}{p} &= 0 , \\ \underset{\sim}{q}^T \underset{\sim}{F}_s &= 0 . \end{aligned}\right\} \tag{I}$$

Since the rank of $\underset{\sim}{F}_s$ is assumed to be $L - 1$, it is possible to normalize $\underset{\sim}{q}$ and $\underset{\sim}{p}$,

$$\underset{\sim}{q}^T \underset{\sim}{p} = 1 . \tag{J}$$

We define a new matrix $\underset{\sim}{H}$ by

$$\underset{\sim}{H} = -\underset{\sim}{F}_s + \underset{\sim}{p} ; \underset{\sim}{q}^T , \tag{K}$$

where $\underset{\sim}{p} ; \underset{\sim}{q}^T$ denotes the dyadic product of the vectors $\underset{\sim}{p}$ and $\underset{\sim}{q}^T$

$$\underset{\sim}{p} ; \underset{\sim}{q}^T = \begin{pmatrix} p_1 q_1 & p_1 q_2 & \cdots & p_1 q_L \\ p_2 q_1 & & & \vdots \\ \vdots & & & \\ p_L q_1 & & \cdots & p_L q_L \end{pmatrix}$$

The matrix $\underset{\sim}{H}$ is nonsingular. We have

and

$$\left.\begin{array}{l} \underset{\sim}{H}\,\underset{\sim}{p} = \underset{\sim}{p}\ ,\ \underset{\sim}{q}^T \underset{\sim}{H} = \underset{\sim}{q}^T \\ \underset{\sim}{p} = \underset{\sim}{H}^{-1}\underset{\sim}{p}\ ,\ \underset{\sim}{q}^T \underset{\sim}{H}^{-1} = \underset{\sim}{q}^T . \end{array}\right\} \qquad \text{(L)}$$

We split off those terms of $\underset{\sim}{r}_s(\underset{\sim}{\beta}, \varepsilon)$ which are linear with respect to $\underset{\sim}{\beta}$:

$$\underset{\sim}{r}_s(\underset{\sim}{\beta}, \varepsilon) = \underset{\sim}{F}_s \underset{\sim}{\beta} + \underset{\sim}{h}(\underset{\sim}{\beta}, \varepsilon) \qquad \text{(M)}$$

Because of (H) we have

$$\underset{\sim}{h}(0,0) = 0$$

Let us assume that $\underset{\sim}{h}(\underset{\sim}{\beta}, \varepsilon)$ is sufficiently smooth to permit the following expansions. We add $\underset{\sim}{H}\,\underset{\sim}{\beta}$ on both sides of (G) and obtain

$$\underset{\sim}{H}\,\underset{\sim}{\beta} = \underset{\sim}{H}\,\underset{\sim}{\beta} + \underset{\sim}{r}_s(\underset{\sim}{\beta}, \varepsilon) .$$

Applying (M) and (K) we get

$$\underset{\sim}{H}\,\underset{\sim}{\beta} = \underset{\sim}{p} ; \underset{\sim}{q}^T \underset{\sim}{\beta} + \underset{\sim}{h}(\beta, \varepsilon) ,$$

and because of (K) we have

$$\underset{\sim}{\beta} = \underset{\sim}{p} ; \underset{\sim}{q}^T \underset{\sim}{\beta} + \underset{\sim}{H}^{-1} \underset{\sim}{h}(\underset{\sim}{\beta}, \varepsilon) \qquad \text{(N)}$$

This is just another form of the equation (G) which is an equation for $\beta(\varepsilon)$. But from (N) we shall be able to derive a bifurcation equation which yields the information about the mani fold of the solutions in the vicinity of $e = e_s$.

18. 3 Bifurcation equation

We introduce an (artificial) parameter $\varkappa$,

$$\varkappa = q^T \beta , \tag{O}$$

into (M):

$$\beta = p \varkappa + H^{-1} h(\beta, \varepsilon) \tag{P}$$

(P) is an implicit equation for $\beta(\varkappa, \varepsilon)$. For small values $\varkappa$, ε the vector $\beta(\varkappa, \varepsilon)$ can be obtained from (P) by the method of successive approximations :

$$\beta^{(1)} = \varkappa p + H^{-1} h(0, \varepsilon), \ldots, \beta^{(k+1)} = \varkappa p + H^{-1} h(\beta^{(k)}, \varepsilon), \cdots . \tag{Q}$$

Let $\beta^*(\varkappa, \varepsilon)$ be an (approximate) solution of (P), e. g., given in the form of a series expansion with respect to $\varkappa$ and ε Putting β^* into (P) and multiplying that equation by q^T we obtain, since $\varkappa$ has to satisfy the relation (O),

$$q^T \beta^* = \varkappa = \varkappa + q^T h(\beta^*(\varkappa, \varepsilon), \varepsilon) , \tag{R}$$

or

$$B(\varkappa, \varepsilon) = \underset{\sim}{q}^T \underset{\sim}{h}\left(\underset{\sim}{\beta}^*(\varkappa, \varepsilon), \varepsilon\right) = 0.$$

We have to solve $B(\varkappa, \varepsilon) = 0$ for $\varkappa(\varepsilon)$ Substituting $\varkappa(\varepsilon)$ into $\underset{\sim}{\beta}^*(\varkappa, \varepsilon)$ we get

$$\underset{\sim}{\beta}(\varepsilon) = \underset{\sim}{\beta}^*(\varkappa(\varepsilon), \varepsilon). \tag{S}$$

The equation

$$B(\varkappa, \varepsilon) = 0 \tag{T}$$

has certainly the solution $\varkappa = 0$, $\varepsilon = 0$. But, in general, the solutions $\varkappa(\varepsilon)$ of (T) are not unique in the neighborhood of (0, 0), $\varkappa(\varepsilon)$ can have multiple (real!!) branches to each of which belongs a $\underset{\sim}{\beta}(\varepsilon) = \underset{\sim}{\beta}^*(\varkappa(\varepsilon), \varkappa)$ Thus, the solutions of $B(\varkappa, \varepsilon) = 0$ yield the required information about the bifurcations, (T) is termed bifurcation equation.

18.4 Numerical procedures

18.41 Computation of e_s

In general, it is difficult to compute the value $e = e_s$ for which det $\underset{\sim}{F}(e) = 0$ since Newton's method (to find $\underset{\sim}{b}(e)$, cf. 15.42) does not work for $e \approx e_s$. Thus, the simplified Newton's method or a generalized form of the regula falsi must be applied to get $\underset{\sim}{b}(e)$. If $\underset{\sim}{b}(e)$ is known, e_s can be computed from det $\underset{\sim}{F}(e_s) = 0$ rather easily.

In special cases no serious difficulties arise. Let a system have odd harmonic solutions, say, within a certain parameter range (odd harmonic: the Fourier series contains only terms ω, 3ω, 5ω, ...). Then $\underset{\sim}{b}_0(e)$ can be calculated, $\det \underset{\sim}{F}_0(e) \neq 0$, say. (The subscript 0 indicates that the vector $\underset{\sim}{b}_0$ contains only elements which belong to odd harmonic Fourier terms.)

Let us assume that we are interested in a singular point $e = e_s$ where a non-odd harmonic solution branches off from the known solution. We introduce a more general vector $\underset{\sim}{b}_G$ which contains the elements of the odd harmonic vector $\underset{\sim}{b}_0$ but additionally elements for even harmonic Fourier terms $\underset{\sim}{b}_E$,

$$\underset{\sim}{b}_G = \begin{pmatrix} \underset{\sim}{b}_0 \\ \underset{\sim}{b}_E \end{pmatrix} .$$

Along the original solution $\underset{\sim}{b}_G(e)$ is known

$$\underset{\sim}{b}_G(e) = \begin{pmatrix} \underset{\sim}{b}_0(e) \\ 0 \end{pmatrix}, \qquad 0 \text{ - zero vector.}$$

Thus, $e = e_s$ can be computed from $\det \underset{\sim}{F}_G(e_s) = 0$.

18.42 Computation of $\underset{\sim}{p}$, $\underset{\sim}{q}$ and $\underset{\sim}{H}$

$\underset{\sim}{p}$ and $\underset{\sim}{q}$ have to be calculated from (I) and to be normalized corresponding to (J). If $\underset{\sim}{F}_s$ is a symmetric matrix then $\underset{\sim}{p} = \underset{\sim}{q}$. Numerical difficulties may arise since e_s is not known accurately and det $\underset{\sim}{F}(e_s)$ will differ slightly from zero. Such a trouble can be avoided in the following way Since det $\underset{\sim}{F}$ is computed by means of a Gauss algorithm $\underset{\sim}{F}$ will be transformed into a triangular form in any case. Thus for $e = e_s$ the (single) term in the last line of the triangular form has to vanish. Taking that into account $\underset{\sim}{p}$ and $\underset{\sim}{q}$ can be computed from the triangular form of $\underset{\sim}{F}(e_s)$ quite accurately.

If $\underset{\sim}{p}$ and $\underset{\sim}{q}$ are known, $\underset{\sim}{H}$ can be computed from (K).

18.43 Generation of the bifurcation equation

To obtain the bifurcation equation (T) we have to calculate $\underset{\sim}{\beta}^*(\varkappa, \varepsilon)$ from equation (Q). First, we have to expand the operator D from section 15.4, cf. equation (A) in section 15.41 into a series with respect to u and e (e.g., $e = \omega^2$). Substituting a few terms of that expansion into (Q), cf. equation (A), it is possible to obtain numerically a few terms

$$\underset{\sim}{\beta}^*(\varkappa,\varepsilon) = \varkappa \underset{\sim}{p} + \underset{\sim}{v}_{01}\varepsilon + \underset{\sim}{v}_{20}\varkappa^2 + \underset{\sim}{v}_{11}\varkappa\varepsilon + \underset{\sim}{v}_{02}\varepsilon^2 + \cdots . \qquad \text{(U)}$$

The $\underset{\sim}{v}_{jk}$ represent numerical vectors. Putting (U) into (R) we obtain

$$B(\varkappa,\varepsilon) = A_{01}\varepsilon + A_{20}\varkappa^2 + A_{11}\varkappa\varepsilon + A_{02}\varepsilon^2 \ldots = 0 \qquad \text{(V)}$$

where the A_{jk} are real numbers.

18.44 Solution of the bifurcation equation

Equation (V) represents a truncated series for $B(\varkappa,\varepsilon)$. To find the various real solutions $\varkappa(\varepsilon)$ of the algebraic relation (V) we can apply a graphical procedure. "Newton's polygon", outlined, e.g., in A. R. Forsyth, Theory of functions of a complex variable, 3. ed., Cambridge University Press, Cambridge 1918. On the other hand it is possible to choose numerical values ε, say, and to solve the resulting algebraic equation on a digital computer. (Remark: If $A_{01} \neq 0$, (V) has a unique solution $\varepsilon(\varkappa)$, some of the $b_j(e)$ have a vertical tangent at $e = e_s$, cf. Fig. 18.1, there exist no bifurcations).

After the various $\varkappa(\varepsilon)$ are known, the corresponding $\underset{\sim}{\mathfrak{b}}(\varepsilon)$ can be calculated from (U), we obtain "line elements" of the branching off solutions at the singular point which may be used as starting approximations for the numerical procedure outlined in section 15.4.

Contents

Printed in the United States
By Bookmasters